Himalaya, Dynamics of a Giant 3

In memory of distinguished professors
M. Gaetani, P. Molnar and A. Steck

SCIENCES

Geoscience, Field Director – Yves Lagabrielle

Dynamics of the Continental Lithosphere,
Subject Head – Sylvie Leroy

Himalaya, Dynamics of a Giant 3

Current Activity of the Himalayan Range

Coordinated by
Rodolphe Cattin
Jean-Luc Epard

ISTE

WILEY

First published 2023 in Great Britain and the United States by ISTE Ltd and John Wiley & Sons, Inc.

ISTE Ltd
27-37 St George's Road
London SW19 4EU
UK

www.iste.co.uk

John Wiley & Sons, Inc.
111 River Street
Hoboken, NJ 07030
USA

www.wiley.com

Library of Congress Control Number: 2022948342

British Library Cataloguing-in-Publication Data
A CIP record for this book is available from the British Library
ISBN 978-1-78945-131-3

ERC code:
PE10 Earth System Science
PE10_5 Geology, tectonics, volcanology

Contents

Tributes

In the past five years, three of our friends, colleagues and mentors Maurizio Gaetani, Peter Molnar and Albrecht Steck have passed away. Their outstanding contribution to the knowledge of the Himalayan range has influenced many authors of this book. We pay tribute to them in the following paragraphs. We dedicate this book to these three exceptional professors.

The mountains of Asia, and the charm of romantic geology – A tribute to the legacy of Maurizio Gaetani (1940–2017) by Eduardo Garzanti

Student of Ardito Desio, organizer of the 1954 Italian conquest of K2 and younger colleague of Riccardo Assereto, killed by a landslide during the second Friuli earthquake of September 9, 1976, the everlasting love of Maurizio Gaetani for Asian geology began in 1962, with his Thesis fieldwork in the Alborz Mountains of Iran. During summer 1977, as Ladakh opened to foreigners, Maurizio first discovered with Alda Nicora the Cretaceous/Triassic boundary in the Zanskar Range. On August 1, 1981, we took a bus from Delhi to Lahul and crossed with horses and horsemen the Baralacha La and Phirtse La to describe the stratigraphy of the Paleozoic–Eocene succession of the Tethys Himalaya. New expeditions were led by Maurizio to Zanskar in 1984 and 1987 to reconstruct the paleogeographic history of northern India, from the newly identified Early Paleozoic Pan-African orogeny to Upper Paleozoic Neotethyan rifting and the subsequent Mesozoic passive-margin evolution terminated with early Paleogene collision with the Asian arc-trench system.

In the meanwhile, Maurizio's Karakorum adventure had begun with the 1986 expedition to the Hunza Valley. From then on, Maurizio's unique

Himalaya, Dynamics of a Giant 3,
coordinated by Rodolphe CATTIN and Jean-Luc EPARD.

dedication to Karakorum geology is testified by 10 expeditions he led to Chitral, Wakhan, Shimshal and Shaksgam, during which every meter of the stratigraphic section from Ordovician to Cretaceous was measured to reconstruct the opening of Paleotethys and Neotethys and their subsequent closure during early and late Mesozoic orogenies. The amazing amount of work carried out during these surveys is condensed in the magnificent geological map of North Karakorum and summarized in his last paper "Blank on the Geological Map". The legacy of Maurizio Gaetani is to remind us that, no matter how much technology is involved, any scientific adventure is primarily a romantic adventure.

Maurizio Gaetani has contributed to about 50 articles published in peer review journals. Here is a list of his major contributions:

Gaetani, M. (2016). Blank on the geological map. *Rendiconti Lincei*, 27(2), 181–195.

Gaetani, M. and Garzanti, E. (1991). Multicyclic history of the northern India continental margin (northwestern Himalaya). *AAPG Bulletin*, 75(9), 427–1446.

Gaetani, M., Nicora, A., Premoli Silva, I. (1980). Uppermost Cretaceous and Paleocene in the Zanskar range (Ladakh-Himalaya). *Rivista Italiana di Paleontologia e Stratigrafia*, 86(1), 127–166.

Gaetani, M., Garzanti, E., Jadoul, F., Nicora, A., Tintori, A., Pasini, M., Khan, K.S.A. (1990). The north Karakorum side of the Central Asia geopuzzle. *Geological Society of America Bulletin*, 102(1), 54–62.

Zanchi, A. and Gaetani, M. (2011). The geology of the Karakoram range, Pakistan: The new 1: 100,000 geological map of Central-Western Karakoram. *Italian Journal of Geosciences*, 130(2), 161–262.

Always on the cutting edge and looking for new ideas to advance Earth Sciences – Peter Molnar (1943–2022) by Vincent Godard, Rodolphe Cattin and György Hetényi

In the final phase of preparing the three volumes of this book, we sadly learned of the passing of Peter Molnar. Peter was a giant of the Earth Sciences, whose contributions would be too long to list exhaustively in this tribute; he left a remarkably enduring mark on Himalaya–Tibet research.

Peter's scientific career began at a pivotal moment for Earth Sciences. He has significantly contributed to developing and applying the new paradigm of Plate Tectonics. Using innovative approaches based on tectonics, seismology, paleomagnetism and satellite imagery, Peter revolutionized our understanding of continental deformation and lithosphere behavior. Although Peter worked on a wide range of problems and geoscientific contexts, the India–Asia collision and the dynamics of the Himalaya–Tibet system have very often been at the core of his investigations. He has left a lasting imprint on research in this region, particularly by his desire to understand the mechanical processes at work during the deformation of this orogenic system.

Consistent with his comprehensive and integrated approach to geodynamic problems, Peter explored a wide range of ideas and processes related to the Himalaya–Tibet system's global role, and initiated several ideas and research directions that are still active today. One example, among many, is the study of the physical relationships and interaction of mechanisms between the development of Tibet's topography and the Southeast Asian monsoon regime. Among the research fields initiated by Peter, understanding the relationships between erosion, tectonics and climate is undoubtedly one of the most innovative and impactful for our community. Following a series of seminal articles by Peter and his colleagues, the complex interactions between the processes responsible for topographic relief creation and destruction are still actively debated in the Himalaya. Peter's research always focused on a global understanding of these processes, particularly those responsible for the variations in erosion and global sedimentary fluxes related to the evolution of Himalayan orogeny and the late Cenozoic evolution of climate. The fact that so many of these topics are still at the forefront of current research by so many groups worldwide is a major testimony to Peter Molnar's prescience on the dynamics of the Himalaya–Tibet system.

Beyond these outstanding contributions, Peter will be remembered for his great sense of humor and for having always been accessible and available to discuss new ideas with young scientists.

Peter Molnar has contributed to an impressive number of publications, some of which have been milestones in the understanding of the Tibet–Himalaya system:

Bilham, R., Gaur, V.K., Molnar, P. (2001). Himalayan seismic hazard. *Science*, 293(5534), 1442–1444.

Gan, W., Molnar, P., Zhang, P., Xiao, G., Liang, S., Zhang, K., Li, Z., Xu, K., Zhang, L. (2022). Initiation of clockwise rotation and eastward transport of southeastern Tibet inferred from deflected fault traces and GPS observations. *Geological Society of America Bulletin*, 134(5–6), 1129–1142.

Houseman, G.A., McKenzie, D.P., Molnar, P. (1981). Convective instability of a thickened boundary layer and its relevance for the thermal evolution of continental convergent belts. *Journal of Geophysical Research: Solid Earth*, 86(B7), 6115–6132.

Molnar, P. (2012). Isostasy can't be ignored. *Nature Geoscience*, 5(2), 83–83.

Molnar, P. and England, P. (1990). Late Cenozoic uplift of mountain ranges and global climate change: Chicken or egg? *Nature*, 346(6279), 29–34.

Molnar, P. and Tapponnier, P. (1975). Cenozoic Tectonics of Asia: Effects of a continental collision: Features of recent continental tectonics in Asia can be interpreted as results of the India-Eurasia collision. *Science*, 189(4201), 419–426.

Zhang, P.Z., Shen, Z., Wang, M., Gan, W., Bürgmann, R., Molnar, P., Wang, Q., Niu, Z., Sun, J., Wu, J. et al. (2004). Continuous deformation of the Tibetan Plateau from global positioning system data. *Geology*, 32(9), 809–812.

Geology of the Indian Himalaya – Albrecht Steck (1935–2021) by Jean-Luc Epard and Martin Robyr

Albrecht Steck was the main driving force behind the Himalayan geological research program led at the University of Lausanne over the last 40 years. His work has focused on the Indian Himalaya, particularly on the Mandi to Leh transect. He has directed or supervised eight doctoral theses distributed along this transect. One of the mottos of Albrecht was that any good geological work always starts with a sound geological mapping. Whether in the Alps or in the Himalaya, Albrecht's work excels by the quality of his geological maps. It is becoming a hallmark of Albrecht's work. Indeed, the numerous field missions he led or supervised in Lahul, Zanskar and Ladakh regions allowed the achievement of detailed geological maps covering a remarkable large area of the NW Indian Himalaya.

The research of Albrecht Steck is characterized by the combination of field observations (mapping, stratigraphy, structures and metamorphism) in order to decipher the geometry, kinematics and tectono-metamorphic history associated with orogenic processes. His scientific approach combining a variety of field, structural, petrographic and analytical methods is a hallmark of Albrecht's research. Two publications reflect particularly well the research works led by Albrecht Steck. The first (Steck et al. 1993) concerns a transect from the High Himalayan Crystalline of Lahul to the south to the Indus suture zone to the north; the second (Steck et al. 1998) focused on a complete geological transect through the Tethys Himalaya and the Tso Morari area. The results of these expeditions are also synthesized in a general publication of the Geology of NW Himalaya (Steck 2003). For Albrecht Steck, geology must be made in the field and out of the main touristic roads. His long-term commitment to the detailed study of the NW part of the Himalaya of India significantly contributed to a better understanding of the geology of this region. With Albrecht Steck, the Alpine and Himalayan geological community has lost an eminent researcher and a true Nature lover.

Albrecht Steck has contributed to many geological maps and articles published in peer-reviewed journals. Here are three of his major contributions:

Steck, A. (2003). Geology of the NW Indian Himalaya. *Eclogae Geologicae Helvetiae*, 96, 147–196.

Steck, A., Spring, L., Vannay, J.-C., Masson, H., Stutz, E., Bucher, H., Marchant, R., Tièche, J.-C. (1993). Geological transect across the Northwestern Himalaya in eastern Ladakh and Lahul (A model for the continental collision of India and Asia). *Eclogae Geologicae Helvetiae*, 86(1), 219–263.

Steck, A., Epard, J.-L., Vannay, J.-C., Hunziker, J., Girard, M., Morard, A., Robyr, M. (1998). Geological transect across the Tso Morari and Spiti areas: The nappe structures of the Tethys Himalaya. *Eclogae Geologicae Helvetiae*, 91, 103–121.

Foreword

Rodolphe CATTIN[1] and Jean-Luc EPARD[2]

[1] *University of Montpellier, France*
[2] *University of Lausanne, Switzerland*

The Himalaya is well known as the largest and highest mountain belt on Earth, stretching 2,500 km from the Nanga Parbat syntaxis in the northwest to the Namche Barwa syntaxis in the southeast, with peaks exceeding 8,000 m in altitude. Resulted from the ongoing collision between the India and Asia plates, the Himalaya is frequently used as the type example of a largely cylindrical mountain belt with a remarkable lateral continuity of major faults and tectonic units across strike.

Advances in geoscience over the past few decades have revealed a more complex picture for the dynamic of this giant, with open questions about the initial stages of the Himalayan building, lateral variations in its structures, variations in tectonic forcing, tectonic–climate coupling and assessment of the natural hazards affecting this area.

In this book, we present the current knowledge on the building and present-day behavior of the Himalayan range. The objective is not to be exhaustive, but to give some key elements to better understand the dynamics of this orogenic wedge. The three volumes of this book present (1) the geodynamic framework of the Himalayan range, (2) its main tectonic units and (3) its current activity. The chapters and volumes in this book are self-contained and can be read in any order. However, the three volumes are linked and provide together a self-consistent image of the Himalayan dynamic at various temporal and spatial scales.

Volume 3, entitled "*Current Activity of the Himalayan Range*", addresses the couplings and feedbacks between tectonics, erosion and climate. Natural hazards such as earthquakes, floods and landslides are also presented with a focus on recent scientific findings and societal issues. Finally, three studies illustrate the importance of multidisciplinary approaches to understanding the physical and chemical processes involved in Himalayan dynamics.

This volume is coordinated by Rodolphe Cattin (University of Montpellier, France) and Jean-Luc Epard (University of Lausanne, Switzerland) with the help of the editorial team composed of Laurent Bollinger (French Alternative Energies and Atomic Energy Commission, France), György Hetényi (University of Lausanne, Switzerland), Vincent Godard (University of Aix-Marseille, France), Martin Robyr (University of Lausanne, Switzerland) and Julia de Sigoyer (University of Grenoble, France).

All royalties allocated to the authors of this book will be donated to the "Seismology at School" program (see the next pages).

Preface

From Research to Education: The Example of the Seismology at School in Nepal Program

György HETÉNYI[1] and Shiba SUBEDI[1,2]

[1] *Institute of Earth Sciences, University of Lausanne, Switzerland*

[2] *Seismology at School in Nepal, Pokhara, Nepal*

Scientific research aims at observing and understanding processes, and enriching our knowledge. But who is in charge of transferring this knowledge to society, to everyday life? Can we expect any researcher to become a company CEO, an engineer, a policy maker or a teacher? Our answer is no, not necessarily, but efforts can be made in that direction, and there are successful examples.

In the context of Himalayan geoscience research, a tremendous amount of information exists. It cannot be all simplified and all translated to local languages of the Himalaya; nevertheless, we found it essential that such knowledge transfer starts. In the aftermath of the 2015 magnitude 7.8 Gorkha earthquake, through a series of fortunate steps, we found ourselves putting down one of the bricks of knowledge transfer by initiating the Seismology at School in Nepal program. Our primary pathway choice was education: in the short-term, raising earthquake awareness and better preparedness can spread

Himalaya, Dynamics of a Giant 3,
coordinated by Rodolphe CATTIN and Jean-Luc EPARD.

through the students to their families, relatives and acquaintanceship; in the longer-term, it is today's school students who will build the next generation of infrastructures.

The program started following a bottom-up approach, with direct cooperation with local schools in Nepal. This ensured motivated participants and direct feedback on the activities and about the needs. The program stands on two main pillars and is described in detail in Subedi et al. (2020a). First, earthquake-related topics have been synthesized and translated to Nepali, together with a series of hands-on experiments, and the local teachers have been trained so that they can teach these in their classes. Second, we have installed relatively cheap seismometers (RaspberryShake 1D) in local schools, which became part of the classroom activities and also recorded waves from earthquakes. This has sparked interest in schools, and the openly and publicly available waveform data is useful for monitoring and research as well. To more closely link these two, we have written a simple earthquake location tutorial that is feasible with typical school computers in Nepal (Subedi et al. 2021).

The program has started in Nepal in 2018; as of 2019, more than 20 schools and seismometers have been involved in the program, and the number is reaching 40 in 2022. There is measurable improvement in students' knowledge (Subedi et al. 2020b), and the feedbacks are very positive. Parallel to classical educational pathways, a series of other activities have been developed in the Seismology at School in Nepal program. Each school has received an Emergency Meeting Point sign in Nepali language. Over 6,000 stickers reminding about earthquakes have been distributed to increase awareness (see Figure 8.10 in Volume 3 – Chapter 8). An Earthquake Awareness Song has been written and composed, and became popular on YouTube (https://www.youtube.com/watch?v=ymE-lrAK0TI). We studied the Hindu religious representation and traditional beliefs about earthquakes (Subedi and Hetényi 2021). Recently, we have developed an educational card game to improve the practical preparation and reaction to earthquakes. Finally, we maintain a website with all information openly available (http://www.seismoschoolnp.org).

The program has so far run on funding that is considered to be small in the research domain, and this has covered the cost of materials and the work in Nepal. More recently, a crowd-funding campaign has been started and evolved successfully – we are very grateful to all funders and donators! In the future, the program aims at growing further, all across Nepal and hopefully all along the Himalaya. This will require more manpower and more funds. The authors of this book have generously given their consent to transfer all royalties to the Seismology at School in Nepal program thank you very much!

There is a strong similarity between this book and the Seismology at School in Nepal program: they both aim at taking research results and carrying them to non-specialists. This book is planned to be published in several languages, and to reach students and interested people around the world. The educational program in Nepal aims at bringing earthquake knowledge to those who really need it as they live in a high hazard area. Both efforts aim at increasing awareness, and, thereby, we hope and wish that their effects reach further across all society.

October 2022

References

Subedi, S. and Hetényi, G. (2021). The representation of earthquakes in Hindu religion: A literature review to improve educational communications in Nepal. *Front Commun*, 6, 668086. doi:10.3389/fcomm.2021.668086.

Subedi, S., Hetényi, G., Denton, P., Sauron, A. (2020a). Seismology at school in Nepal: A program for educational and citizen seismology through a low-cost seismic network. *Front Earth Sci*, 8, 73. doi:10.3389/feart.2020.00073.

Subedi, S., Hetényi, G., Shackleton, R. (2020b). Impact of an educational program on earthquake awareness and preparedness in Nepal. *Geosci Commun*, 3, 279–290. doi:10.5194/gc-3-279-2020.

Subedi, S., Denton, P., Michailos, K., Hetényi, G. (2021). Making seismology accessible to the public in Nepal: An earthquake location tutorial for education purposes. *Bull Nep Geol Soc*, 38, 149–162.

PART 1

Surface Process

1

Orogenesis and Climate

Frédéric FLUTEAU[1], Delphine TARDIF[2,3], Anta-Clarisse SARR[4], Guillaume LE HIR[1] and Yannick DONNADIEU[4]

[1]Institut de physique du globe de Paris, Paris Cité University, France
[2]LSCE, Gif-sur-Yvette, France
[3]Institut de Recherche pour le Développement, Montpellier, France
[4]CEREGE, Aix-en-Provence, France

1.1. Introduction

Climate is controlled by the incoming solar energy (a quantity that depends on the latitude of the site, the solar constant and the obliquity and climatic precession), the greenhouse effect related to the atmospheric composition and the energy redistribution by atmospheric and oceanic circulation. Climate also depends on the exchanges (energy, matter and momentum) between the components of the climate system: atmosphere, ocean, cryosphere, biosphere, lithospheric surface. A disturbance applied to one or more of these components leads to a destabilization of the system, resulting in climate change. Climate changes have various causes and occur at all timescales. Causes can be internal to the climate system and related to the natural variability of physical processes, such as the ocean–atmosphere coupling or the instability of an ice cap, as well as external to the climate system, such as plate tectonics and one of its features, orogeny.

Himalaya, Dynamics of a Giant 3,
coordinated by Rodolphe CATTIN and Jean-Luc EPARD.

In this chapter, we discuss how orogenesis alters the climate system. To do so, we take the example of the uplift of the Himalayan range and Tibetan plateau, as a result of the India–Asia collision (see Volume 1 – Chapter 1). After a short description of the present-day climate in Asia, we describe the climate evolution during the Cenozoic at the global scale and in Asia. We use climate modeling studies to understand the responses of the atmosphere and the ocean to mountain uplift. A climate model is a mathematical representation of the climate system. It simulates the climate at equilibrium with imposed boundary conditions such as paleogeography, atmospheric chemistry, and Earth's orbital parameters. Modifying the boundary conditions thus allows us to quantify the impact of these factors on the climate.

The climatic impact of an orogeny is not limited to the response of the atmospheric and oceanic circulation, but also induces an indirect response of the climate system by affecting the carbon cycle. Indeed, changes in temperature and precipitation that may occur as a result of orogeny will in turn affect the physical erosion of landforms (see Volume 3 – Chapter 2). Additionally, sediments can undergo chemical erosion that will further disrupt the carbon cycle and therefore the climate. Other mechanisms are also likely to act on the carbon cycle, such as the burial of organic matter. This orogeny–carbon cycle–climate relationship has been studied in the case of the Himalayan orogeny.

1.2. Climate in Asia: present and past

1.2.1. *Present-day climate*

Asia has an extremely contrasted climate. A large southern and southeastern area, stretching from Pakistan to eastern China, undergoes the seasonal reversal of monsoon flow (South Asian monsoon from Pakistan to Myanmar and East Asian monsoon from Myanmar to eastern China). A vast region to the west and north of the Tibetan plateau (from Iran to Mongolia) experiences a semi-arid to arid climate, while further north, Siberia is dominated by a continental to polar climate. This dichotomy between the warm and seasonally wet climates of south and eastern Asia and the inland arid belt is due to the presence of high mountain ranges (Himalaya, Karakorum, Pamir, Hindu Kush, Zagros) and the Tibetan plateau, which strongly limits the atmospheric exchanges between the low and mid-latitudes of Eurasia. The Asian monsoon is marked by a strong seasonality of precipitation (with most precipitation occurring during summer), coupled with a reversal of the atmospheric circulation. Halley (1686) described the monsoon as a result of differential heating between the Indian Ocean and the Asian continent. This approach is still widely used (Webster et al. 1998), even though some scientists consider monsoon as the manifestation of the seasonal migration of the intertropical convergence zone.

In winter, in the mid-latitudes of Asia (mostly Siberia), the low insolation and the continentality amplified by the snow albedo-temperature feedback cause extremely low temperatures and generate the Siberian anticyclone, the densest air masses in the Northern Hemisphere, which is maintained by radiative cooling at the surface and large-scale descending motion (Ding and Krishnamurti 1987). A strong temperature gradient between the cold Asian continent and the mild Pacific Ocean results in the advection of cold and dry continental air to the northeastern China. To the south, East Asian winter monsoon generates moderate precipitation over the southeastern China (Figure 1.1) fed by moist air blowing from the nearby Pacific ocean (Figure 1.2).

At the end of the winter season, the increase in insolation over the northern hemisphere favors the rise of surface temperatures, which are also amplified locally by the decrease of albedo due to the melting of snow cover. Over southern Asia, convective systems develop and a low-pressure cell settles over India. At the end of spring, surface heating of a large part of Asia continent leads to the establishment of deep convection. In summer, the peak upper-tropospheric temperatures at 300 hPa (about 9,500 m high) located to the south of the Tibetan plateau are thermodynamically coupled to the precipitating convection (Boos and Kuang 2010). At the surface, a deep low-pressure cell stretches over southern Asia up to the Arabian Peninsula and conveys air masses from the Indian Ocean (Figure 1.2). Despite a moderate mean elevation of about 2,000 m, except in the Ethiopian region, the East African highlands constitute an orographic barrier for the air masses advected by the Indian low pressure. The low-level cross-equatorial flow, referred to as the Somali jet, carries moisture that leads to heavy precipitation over Asia. While reaching India, the humid air mass hits the Great Escarpment of the Deccan Traps, a pile of basaltic lava flows culminating at 1,700 m in the Western Ghats, leading to strong precipitation on the coastal fringe (Figure 1.1), causing recurrent catastrophic floods in the metropolis of Mumbai and its 20 million inhabitants. To the east of the Ghats landform, rainfall is less

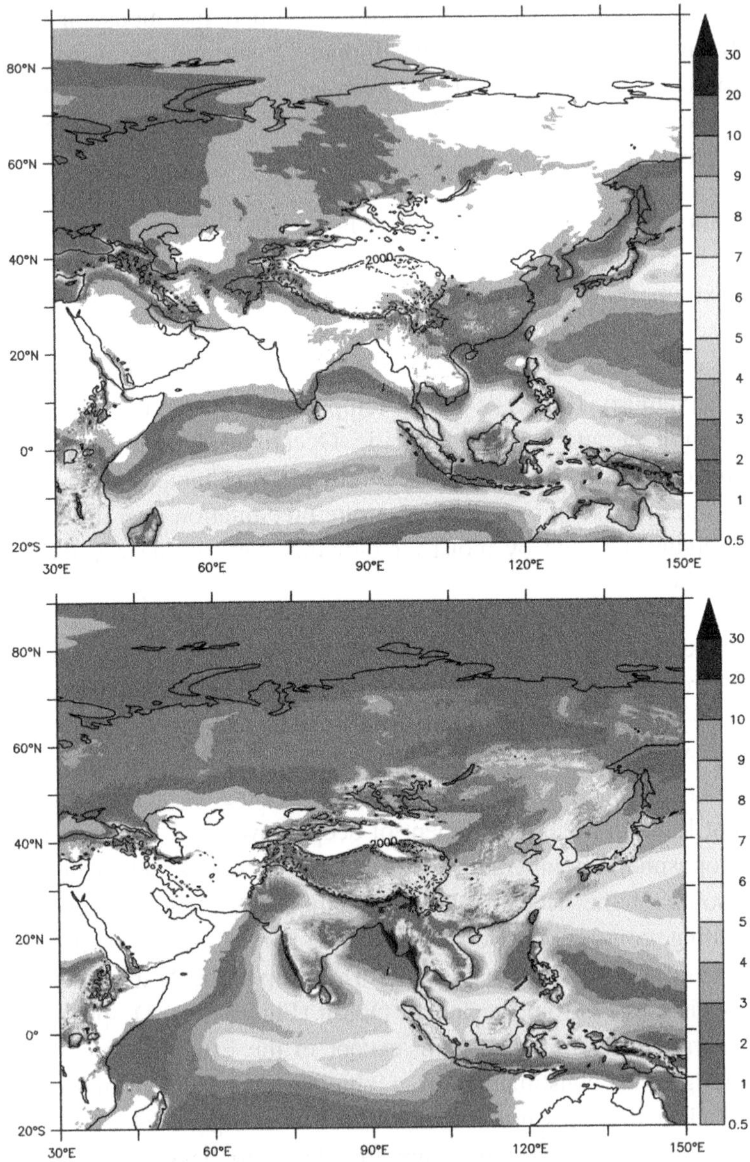

Figure 1.1. *Mean precipitation in winter (top) and summer (bottom) (in mm/day) (data are from ERA5 reanalysis (Hersbach et al. 2019)). Topography contoured at 2,000 m (4,000 m) in a thin solid (dash) line. For a color version of this figure, see www.iste.co.uk/cattin/himalaya3.zip*

intense due to a rain shadow effect. The air masses are reloaded with moisture over the Bay of Bengal, before hitting the Indo-Burmese range to the east and the Himalayan range to the north (Figure 1.2), causing an intense orographic rainfall (Figure 1.1). Indeed, the Himalaya forms an east–west gigantic barrier to the north of the alluvial plains of the Indian subcontinent. As the air masses are lifted on the lee of the Himalaya and in the convective systems, they cool down, thus decreasing the maximum water vapor they can contain, according to the Clausius–Clapeyron relationship. When the saturation vapor pressure is reached, the water vapor condenses on nuclei (mineral dust, salt crystal, volcanic ash) to form water droplets (or ice crystals depending on the air temperature), leading to cloud formation. Water condensation released a large amount of latent heat which self-sustains the monsoon circulation in Asia. Water droplets grow by coalescence in the clouds and turn to heavy rainfalls. The eastern part of the Tibetan plateau also experiences strong summer precipitation, while the western part of the plateau is drier (Figure 1.1).

We will now propose an overview of how the climate has evolved since the India–Asia collision based on paleoclimatic indicators.

1.2.2. *Cenozoic climate evolution*

The Earth's climate during the Cenozoic is marked by a succession of slow changes interspersed with abrupt transitions. After the Early Eocene climatic optimum (~50 Ma), during which the Earth experienced a globally hot climate, a slow cooling is recorded, ending with the formation of the Antarctica ice cap around the Eocene–Oligocene boundary (~34 Ma) (Ladant et al. 2014; Westerhold et al. 2020). This unipolar glaciation persisted through the Oligocene and Early Miocene. The Earth undergoes a new cooling period starting at ~14 Ma, which lead to the final build-up of the Antarctic ice sheet approximately 13–14 Ma followed by the Greenland glaciation at the end of the Miocene (~6 Ma).

Asia has undergone important climatic changes during the Cenozoic that can be deduced from different proxies (lithology, flora, isotopes, etc.). These evidences are numerous and we will focus on a few sites only. During the early Cenozoic (Paleocene–Eocene) at the time of the India–Asia collision, southern Asia experienced a warm and wet climate as evidenced by the fossil flora discovered in western India (Rajasthan, Gujarat) (Spicer et al. 2017; Bathia et al. 2021).

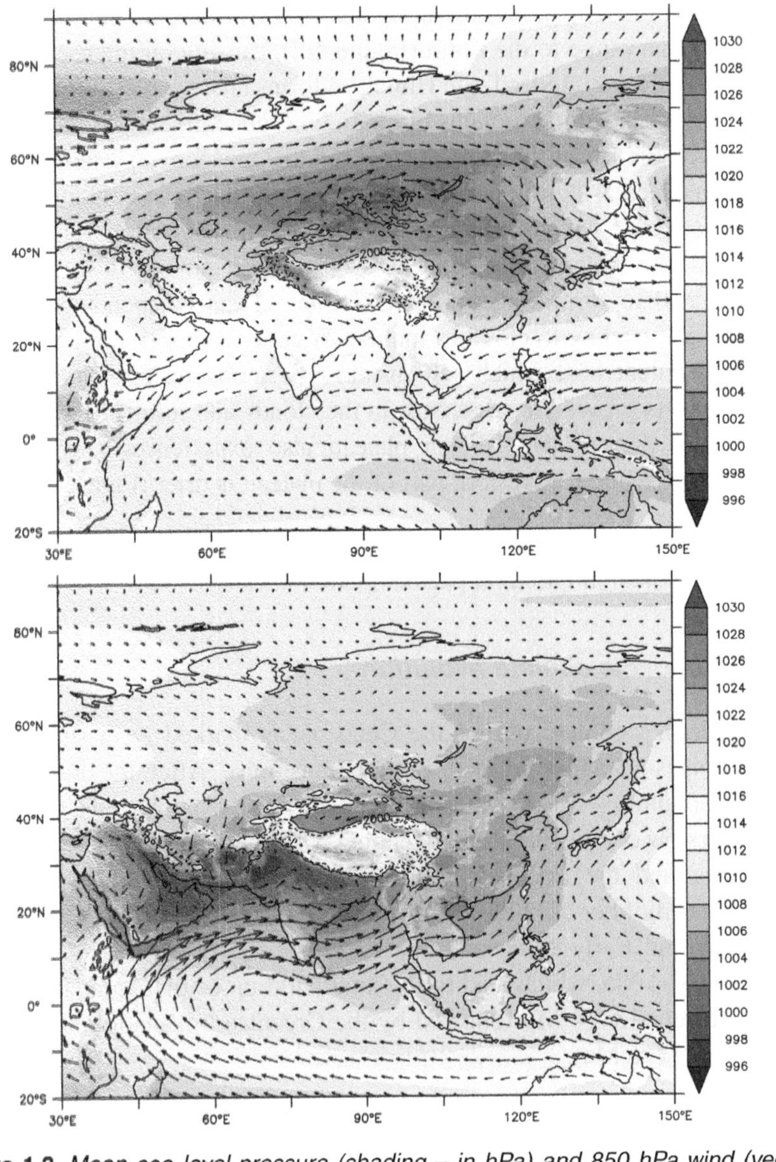

Figure 1.2. *Mean sea level pressure (shading – in hPa) and 850 hPa wind (vector) in winter (top) and summer (bottom) (data are from ERA5 reanalysis (Hersbach et al. 2019)). Light to dark blue colors are for low pressure, yellow to orange colors are for high pressure. Topography contoured at 2,000 m (4,000 m) in a thin solid (dash) line. For a color version of this figure, see www.iste.co.uk/cattin/himalaya3.zip*

In Burma, low oxygen isotope values ($\delta^{18}O$) of freshwater gastropod shells from the Late Eocene indicate a marked seasonality of precipitation (Licht et al. 2014). Farther east, in the Yunnan Basin (China), xerophytic flora and sedimentation consisting of red sandstone and gypsum provide evidence of a warm and semi-arid climate that persisted until the early Late Eocene (~41 Ma) (Fang et al. 2021) followed by a warm and wet climate in the late Eocene (~35 Ma) (Sorrel et al. 2017). In central Tibet, an abundant fossil flora from the middle Eocene (~47 Ma) found in the Bangor Basin reflects a subtropical climate and seasonal rainfall (Su et al. 2020). This floristic assemblage has strong affinities with European and North American flora but very little with Gondwanan flora of India, suggesting the presence of high landform in the south of this basin (Gangdese Uplands) prior to the rise of the Himalaya. All of these sites, except the Yunnan site, record more or less pronounced seasonality of precipitation that has been interpreted as evidence of a monsoonal regime as early as the Eocene (Licht et al. 2014) or the seasonal migration of the intertropical convergence zone (Bathia et al. 2021).

The onset of a modern monsoon regime in Asia is thought to date from the early Miocene based on paleobotanical and lithological data (Sun and Wang 2005). This is coeval with the intensification of chemical weathering of the sediment deposited on the continental margin of the South China sea, as recorded by the K/Al ratio (Clift and Webb 2019). The appearance of upwelling-sensitive planktic foraminifer Globigerina bulloides off the Oman margin driven by the summer monsoon wind is much more recent, ~13 Ma (Gupta et al. 2015). Thus, the age of onset of the modern monsoon regime has long been debated. We will see that this paradox is now resolved (Sarr et al. 2022). During the late Miocene, the intensity of the Asian monsoon decreases in response to global cooling. Oxygen isotope analysis on fossil freshwater bivalve shells and mammal teeth from the Himalayan foreland yields the seasonal $\delta^{18}O$ variation in surface waters that suggests a decline in precipitation approximately 7.5 Ma (Dettman et al. 2001). This agrees well with a diachronous change in vegetation from trees and shrubs (using the C3 photosynthetic pathway) to grassland (using the C4 photosynthetic pathway) found in the Siwaliks in the Himalayan foothill (Quade et al. 1989; Tauxe and Feakins 2020) and ranging from 7.8 Ma in Pakistan to 6 Ma in Nepal, together with the evidence of grassland expansion recorded in the Weihe Basin (central China) (Zhao et al. 2020).

In Central Asia, the climatic evolution is very different. The Asian continental interior becomes more arid from the Late Eocene onward. Eolian sedimentation associated with gypsum deposited in the Xining Basin is representative of playa-lake environment (Meijer et al. 2019) in which a xerophytic fossil flora develops (Barbolini et al. 2020). This trend is reinforced from the Oligocene onward with the gradual disappearance of gypsum in the Xining Basin (Fang et al. 2015). Loess deposits in northwest China dated to the Lower Miocene (~22 Ma) are interpreted as a marker of the winter monsoon (Guo et al. 2008). An arid climate has persisted since the early Miocene (Barbolini et al. 2020).

To understand the climatic impact of paleogeographic changes, climate modeling proves to be a relevant tool. However, this requires first reconstructing the paleo-elevation of landforms.

1.3. Reconstructing the paleo-elevation of landforms

Orogenesis leaves structural and petrological markers that can be used to identify past deformation. However, the main difficulty lies in estimating the paleo-elevation of the landforms. We discuss hereafter the three main methods employed to obtain paleo-elevation estimates.

The relationship linking leaf physiognomy (size, shape, thickness, type of leaf margin) to climate is assumed to have remained constant over time. The analysis of specific morphological features of the fossil leaves therefore makes it possible to estimate the paleo-temperature at one site. Moreover, by comparing the fossil flora to a nearby site of similar age known to be located at sea level, we can deduce the paleo-elevation of the site by fixing the lapse rate (which varies from 4°C/km to 10°C/km) or enthalpy (Spicer et al. 2018). This method is calibrated on dicots and does not take into account all biogeographic regions (such as Australia) or taxa (such as conifers). It cannot be employed before the upper Cretaceous since the flowering plants appear during the Cretaceous.

A second method is based on the oxygen isotope measurement in pedogenic and lacustrine carbonates (Rowley and Garzione 2007). The ratio of heavy to light oxygen isotopes, when compared to a global reference value (namely the Vienna standard mean ocean water, VSMOW), is noted ($\delta^{18}O$). The decrease of $\delta^{18}O$ with altitude reflects a depletion in heavy isotope (^{18}O) of precipitation along the trajectory of the air mass (Rayleigh-type distillation process) (Mulch et al. 2016). Assuming the precipitation-calcite thermodynamic equilibrium, the $\delta^{18}O$ of lacustrine and pedogenic carbonates represents the isotopic signature of rainfalls and thus the paleoaltitude of the site. This tool has, however, some drawbacks: 1) the isotopic data represents surface water catched by the draining basin rather than rainfall and 2) the isotopic signature depends on the air mass pathway. To circumvent this difficulty, a new promising tool referred to as the clumped isotope measurement Δ_{47} has been developed. The clumped isotope measurement relies on the propensity to form molecules containing bonds between heavy isotopes of carbon and of oxygen in a carbonate. The quantity of bonds is a function of the temperature.

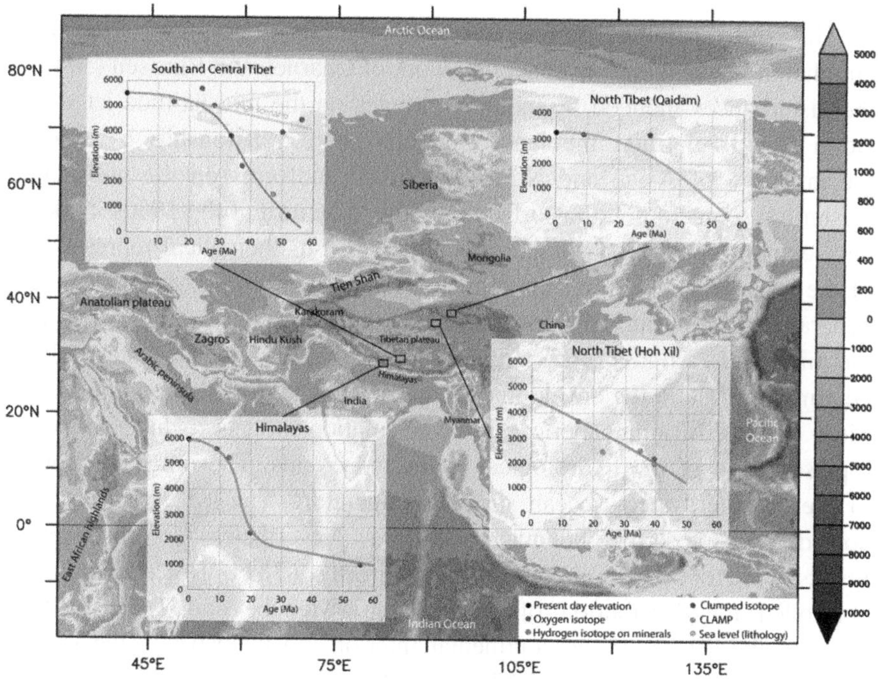

Figure 1.3. *Shading: present-day geography (m). Plots display the paleo-elevation of the Himalaya, Southern and Central Tibet, Hoh Xil basin and Qaidam basins. The filled circles represent different proxies of paleo-elevation. The gray curves are indicative of the elevation during the Cenozoic. For a color version of this figure, see www.iste.co.uk/cattin/himalaya3.zip*

These tools have been employed to reconstruct the paleo-elevation of landforms in Asia (Figure 1.3). The southern and central Tibetan Plateau are thought to have reached a high elevation of ~4 km by the late Eocene (Ding et al. 2017). This scenario, however, remains debated because the $\delta^{18}O$ of rainfall could be biased by climatic processes. Indeed, an increase in convective precipitation or a different air mass trajectory may result in a decrease of $\delta^{18}O$ of rainfall, suggesting a relief higher than it actually was (Botsyun et al. 2019). The Himalayan range reaches its current elevation in the middle Miocene ~15 Ma (Gébelin et al. 2013; Ding et al. 2017). The northern part of Tibet and the ranges north of the Tibetan Plateau (Tien Shan, Altai) mostly rise from the Miocene onward.

1.4. The contribution of climate modeling

The understanding of the effects of landforms on the atmospheric circulation began with the pioneering work of Charney and Eliassen (1949) on the disturbances induced by mountain ranges on the mid-latitude jet stream. Bolin (1950) clarified the landform–atmosphere relationship by studying the influence of Asian and North American landforms on the mid-latitude westerlies, temperature and surface pressure. The turn of the 1960s marks the development of the first climate model that is a mathematical representation of the processes governing the climate system. The first models, also called general circulation models, only simulated the atmosphere circulation. Yet, in the mid-1960s, Mintz (1965) already showed that the Tibetan plateau plays a determining role in the development and location of the Siberian anticyclone. Since the 1970s, atmospheric general circulation models include an explicit representation of the hydrological cycle and clouds. The other components of the climate system (ocean surface temperature, sea ice extent, etc.) are considered as prescribed boundary conditions. Nevertheless, several important results were obtained: the uplift of landforms (Tibet, Rocky Mountains) controlled the distribution of precipitation (Manabe and Terpstra 1974), favoring the drying of the continental interior and intensifying the summer monsoon in Asia (Hahn and Manabe 1975). In the late 1980s, Kutzbach et al. (1989) showed how the mid-latitude upper-tropospheric planetary waves increase with mountain uplift and how low-tropospheric winds are diverted by topography. They emphasized the consistency of the simulated climate changes due to uplifts with late Cenozoic climate indicators (Ruddiman and Kutzbach 1989). Until the early 1990s, studies were conducted by modifying the height of the landform on a present-day geography. Subsequent work considers late Cenozoic paleogeographic changes to assess the impact of landform changes (Fluteau et al. 1999).

More sophisticated climate models referred to as Earth system models were developed in the early 2000s and have benefited from the development of high-performance computing. These climate models simulate the atmosphere, the ocean, the sea ice, the land surface and vegetation and the biogeochemistry of the ocean. The atmospheric and oceanic components are described by a set of equations representing dynamic processes (differential equations of fluid motion calculated on a 3D grid) and physical processes (calculated on a single atmospheric column such as radiative balance, clouds, precipitation, exchanges with other components of the climate system). The land surface model simulates the vegetation, soil and water routing (McGuffie and Henderson 2005; Sepulchre et al. 2020). Because of their very long response time (>10^3 years), ice-sheets are prescribed in the same way as paleogeography, solar constant, atmospheric chemistry and Earth's orbital parameters. Sensitivity experiments to paleogeographical changes allow us to simulate the consequences of uplift on each component of the climate system. Finally, model-data comparison remains an essential step to test the validity of these scenarios.

1.4.1. *Impact of orogenesis on the atmospheric circulation*

The strengthening of the summer monsoon in Asia has regularly been attributed to the uplift of the Tibetan plateau alone, as the heating of its surface favors the convergence of air masses (Kutzbach et al. 1989). However, these early studies did not separate the uplift of the Tibetan plateau from that of the Himalaya. In fact, it has been shown that the South Asian summer monsoon is little impacted when the Tibetan plateau is suppressed and the Himalaya kept at their current elevation (Boos and Kuang 2010; Acosta and Huber 2020). Indeed, a deep convection is set up over northern India leading to the development of a low-pressure area over the Gangetic plain driving the advection of moist air masses from the Indian Ocean. Arriving on the Himalaya, moisture flux condenses and produces orographic precipitation. Maffre et al. (2018) showed that the monsoon in Asia weakens considerably when global landforms are lowered down to 200 m. The low pressure located over northern India and responsible for the advection of moist air mass from the Indian Ocean is replaced by a huge low-pressure area centered in Mongolia stretching over a large part of Eurasia. It results in a decrease of 60–80% of precipitation in the South Asian monsoon domain.

The uplift of Tibet also acts on the circulation of air masses at mid-latitudes, and in particular on the location of the jet stream. With a high Tibetan plateau (>3,000 m), Zoura et al. (2019) shows that the jet stream migrates further north in summer, which increases precipitation in spring (pre-monsoon) and summer (monsoon) in eastern China. Central Asia experiences an aridification when the Tibetan plateau and adjacent landforms are uplifted, which largely explains the observed climatic evolution in this region since the Eocene.

Most of the early work to understand the interaction between topography and the atmospheric circulation was based on sensitivity experiments to the landforms (Tibet, Himalaya) in a modern geographical context. To understand the evolution of the Cenozoic climate, all paleogeographic changes (continental drift, evolution of oceanic basins and epicontinental seas as well as evolution of all landforms) must be taken into account. Tardif et al. (2020) recently showed that in the Late Eocene, the monsoon regime over southern Asia is driven by the seasonal migration of the intertropical convergence zone (Figure 1.4).

The paleogeography at that time is very different from that of the present day with lower landforms, a huge epicontinental sea in Eurasia – the Paratethys – and a largely flooded Arabian Peninsula.

The simulated climate is generally in good agreement with the data except for the Tibetan plateau which is too arid. The onset of a monsoon regime similar to the present-day one in Asia occurs in experiment with the Early Miocene boundary conditions in response to the important paleogeographic changes in Asia (uplift of the Himalaya and adjacent landforms and final retreat of the Paratethys Sea) (Figure 1.5).

Precipitation strengthens over southern Asia (Figure 1.5), especially over the Himalayan range (Sarr et al. 2022) and eastern China (Farnsworth et al. 2019).

However, summer winds in the Arabian Sea region are still too weak to trigger ocean upwelling. The increase in land exposure in the Arabian Peninsula region combined with the uplift of the East African highlands (Africa), Zagros range and Anatolian Plateau (Middle-East) allow us to strengthen the Somali Jet and promote the development of oceanic upwelling in the Arabian Sea during the Late Miocene (Figure 1.6, Sarr et al. 2022).

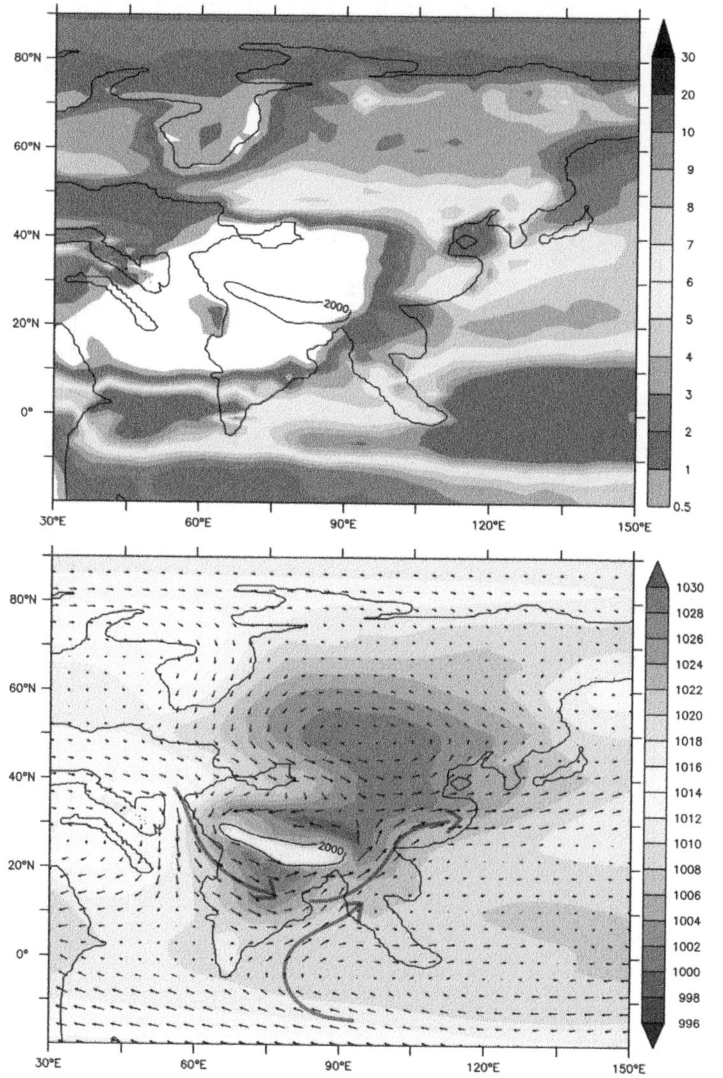

Figure 1.4. *Late Eocene (40Ma) simulated climate with the ESM IPSL-CM5A2. Upper: mean summer precipitation (top – in mm/day). Lower: mean sea level pressure (shading – in hPa) and 850 hPa wind (vectors) in summer. Topography contoured at 2,000 m (4,000 m) in a thin solid (dash) line. The pCO_2 is fixed at 1,160 ppmv. For a color version of this figure, see www.iste.co.uk/cattin/himalaya3.zip*

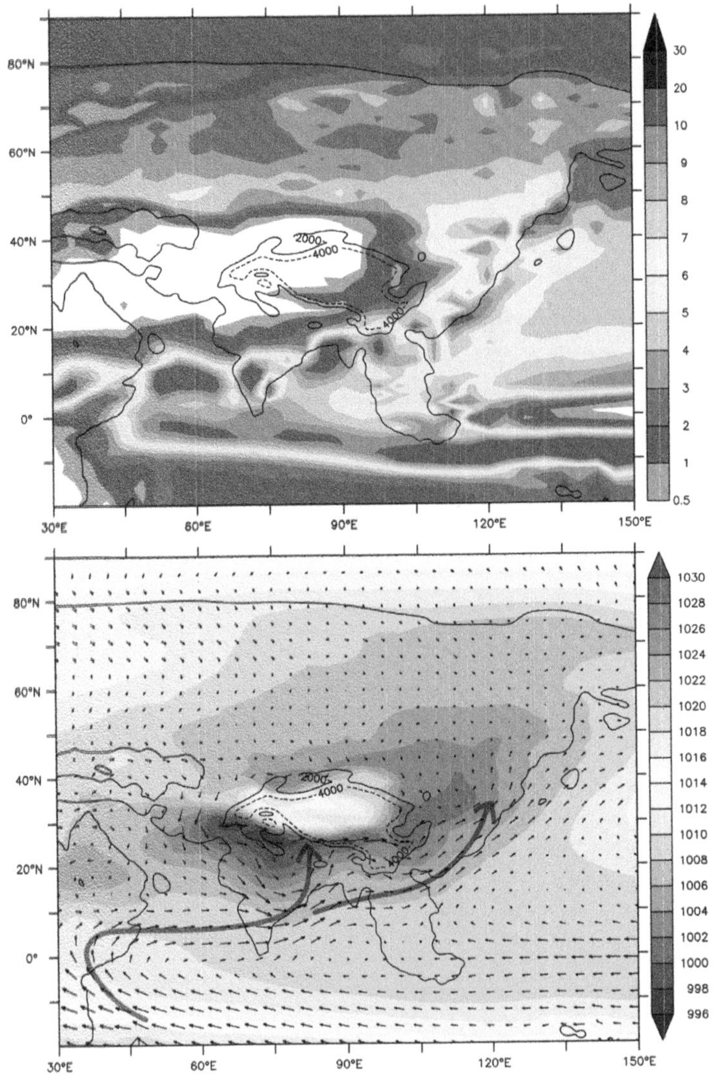

Figure 1.5. *Early Miocene (20 Ma) simulated climate with the ESM IPSL-CM5A2. Same as Figure 1.4. The* pCO_2 *is fixed at 560 ppmv. For a color version of this figure, see www.iste.co.uk/cattin/himalaya3.zip*

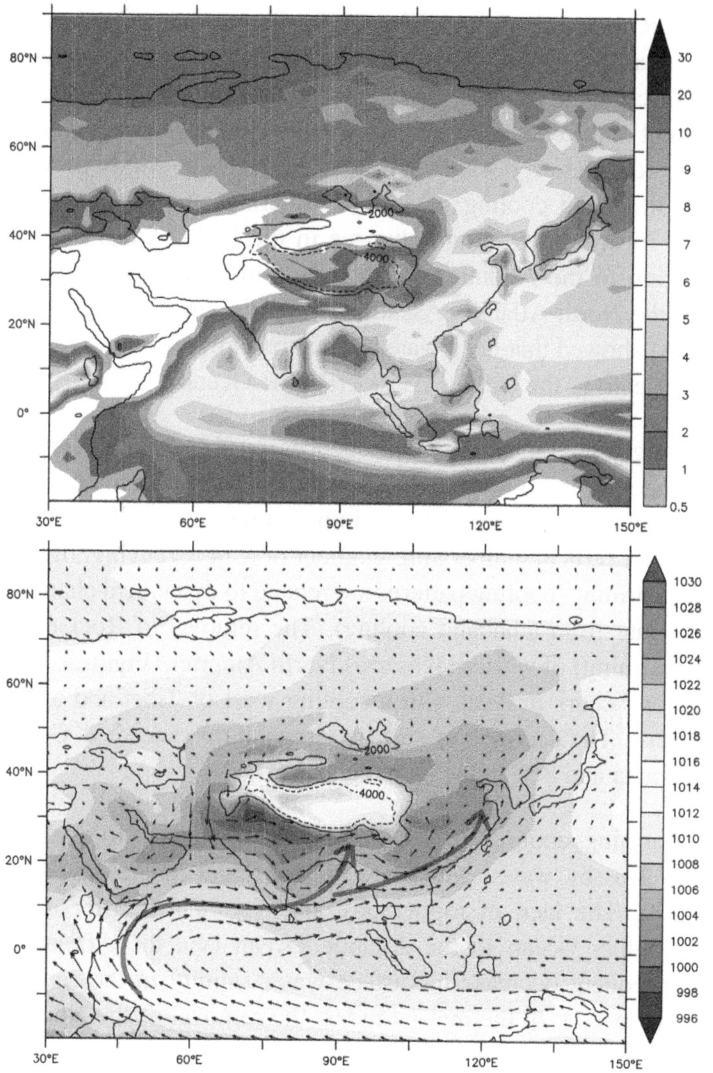

Figure 1.6. *Late Miocene (11 Ma) simulated climate with the ESM IPSL-CM5A2. Same as Figure 1.4. The pCO_2 is fixed at 420 ppmv. For a color version of this figure, see www.iste.co.uk/cattin/himalaya3.zip*

1.4.2. *Impact of orogenesis on the ocean circulation*

Earth System Models also help us to quantify the effects of orogenesis on the ocean circulation. Yang and Wen (2020) showed that when Asian reliefs are removed on a present-day geography, the intensity of the Atlantic meridional overturning circulation (AMOC) is reduced by about 80% in response to the enhancement of the buoyancy of surface waters in the Norwegian Sea. Changes in mid-latitude westerly winds and weakening trade winds over the Pacific Ocean result in the cooling of surface waters in the North Atlantic and a transfer of freshwater (through changes in the hydrological cycle) from the Pacific region to the Atlantic, which results in a decrease of surface water salinity in the North Atlantic. Assuming a nearly flat Earth (<200 m), Maffre et al. (2018) showed that deep water formation shifts from the Norwegian Sea (Atlantic Ocean) to the Gulf of Alaska (Pacific Ocean).

Removing topography has consequences on the surface temperature due to the changes in the atmospheric circulation. The mean annual temperature anomalies (flat Earth experiment minus present-day Earth experiment) corrected by a vertical temperature gradient of 6.14°C/km highlight the region where the warming–cooling patterns are related to dynamical changes (in other words, deviate from adiabatic cooling). The removal of Rockies causes the cooling (warming) of western (eastern) North America (Figure 1.7). The same phenomenon occurs in Asia with a cooling west of Tibet and a warming on the east coast. The decrease of surface temperature exceeds 4°C in the North Atlantic Ocean and 8°C in the Norwegian sea.

The warming–cooling patterns are clearly related to changes in the large-scale Rossby waves which are almost zonal in the case of flat Earth (Figure 1.8). The lack of topography also favors a transfer of freshwater from the Indian Ocean to the Atlantic Ocean (via the African continent) with a consequent decrease in surface water salinity in the central Atlantic Ocean (Figure 1.7). In the mid-latitudes, the removal of the North American Cordillera induces the transfer of freshwater from the Pacific Ocean to the Atlantic Ocean. This results in an increase (decrease) of the surface salinity in the North Pacific (North Atlantic) Ocean (Figure 1.7). Isotopic measurements ($\delta^{18}O$ and $\delta^{13}C$) on benthic foraminifera from the Atlantic Ocean suggest the establishment of the AMOC around the Eocene–Oligocene boundary

(Hutchinson et al. 2021). In addition to the effect of orogenesis, the ocean circulation is obviously influenced by the closure or opening of ocean gateways during the Cenozoic (Drake, Tasman Sea, Greenland–Scotland Ridge, Fram, etc.).

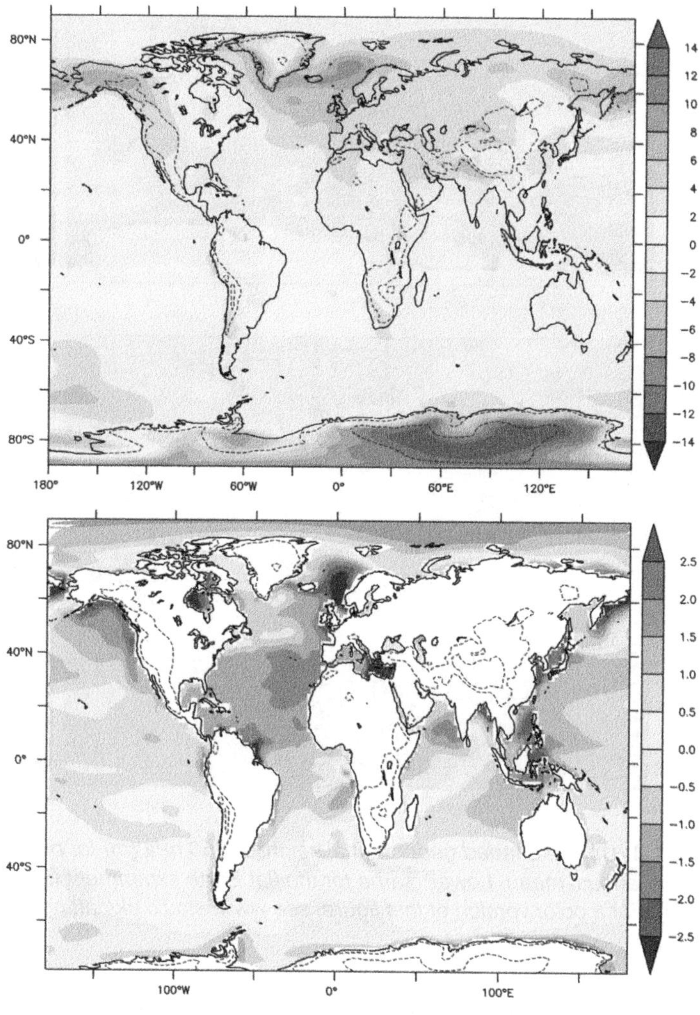

Figure 1.7. *Upper: shaded anomaly (flat Earth minus present-day Earth) of air temperature at 2 m (in °C), annual mean, corrected by a 6.14°C/km lapse rate (after (Maffre et al. 2018)). Lower: shaded anomaly (flat Earth minus present-day Earth) of salt in surface ocean (in psu), annual mean. For a color version of this figure, see www.iste.co.uk/cattin/himalaya3.zip*

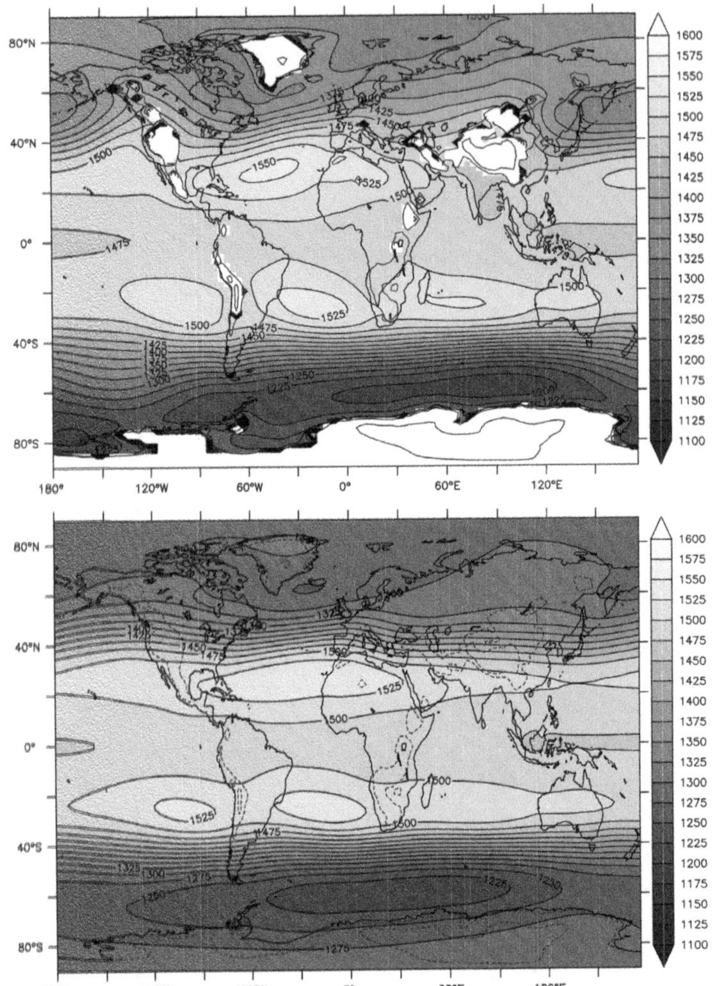

Figure 1.8. *Upper: shaded geopotential height at 850 hPa (m) for present-day simulation, annual mean. Lower: same for the flat Earth experiment (after (Maffre et al. 2018)). For a color version of this figure, see www.iste.co.uk/cattin/himalaya3.zip*

1.4.3. *Impact of orogenesis on the chemical composition of the atmosphere*

In 1899, Chamberlin (1899) suggested in a pioneering work that chemical weathering of silicates associated with mountain uplifts controls climate at a geological timescale. Ninety years later, Raymo et al. (1988) based on ocean records such as the $^{87}Sr/^{86}Sr$ ratio of the ocean concluded that the Himalayan and Tibetan uplifts during the Cenozoic increased physical erosion and enhanced the chemical weathering of sediments with a consequent decrease in the atmospheric CO_2 concentration (pCO_2) and global cooling. While these landforms are all driven by mantle dynamics, their evolution is the result of both endogenous (tectonics) and exogenous (erosion) processes (Champagnac et al. 2012). The erosion rate of a mountain range varies with tectonic activity and climatic context (precipitation and temperature). Fission tracks on apatite collected from the Bay of Bengal sediments trace the exhumation of Himalayan range rocks. However, this rate, estimated at 1–3 km/Ma for the past 13 Ma, appears to be relatively insensitive to climate variations (Huyghe et al. 2020).

Orogenesis obviously has a crucial role as ~40% of chemical weathering is estimated to occur on ~10% of the steepest continental surfaces (Larsen et al. 2014). Yet, carbon consumption by dissolved elements from the Himalayan range (~0.19×10^{12} molC/yr) is only 6% of the total annual flux (3.2×10^{12} molC/yr). The low silicate weathering can be related to a too fast transfer of sediments to the ocean compared to the reaction kinetics of mineral alteration, and the lithology of Himalayan rocks would not be the most favorable (reduced amount of calcium silicate). The consumption of carbon in an orogenic context can take other forms. The transfer and burial of continental organic matter in sedimentary basins has been a very effective mechanism (France-Lanord and Derry 1997). Finally, the orogenic context can be a source of carbon through the oxidation of exhumed organic-rich rocks or sulfide minerals such as pyrite (Hilton and West 2020).

1.5. Conclusion

Climate changes in Asia are largely attributable to paleogeographic changes. The uplifts of the Himalayan range, Tibetan plateau or more modest landforms indeed strongly modify the atmospheric circulation. Associated with other paleogeographic changes (such as the retreat of an epicontinental sea, the uplift of landforms out of Asia or the increase in land exposure in the Arabian Peninsula), the distribution of precipitation has changed spatially and seasonally during the Cenozoic. The uplift of the Himalaya and adjacent landforms (Zagros, Karakorum) progressively reduces the exchanges between mid- and low latitudes and favors the onset of a modern monsoon regime in Asia from the early Miocene, while the uplift of the Tibetan plateau (especially its northern part) disrupts the western moisture flow leading to aridification of Central Asia and to an increase in precipitation in eastern China during spring and summer.

Orogenies cause disturbances in the global ocean by acting on the atmospheric circulation and heat exchanges. One of the consequences of the Cenozoic orogenies is the shift of deep water formation from the Pacific Ocean to the Atlantic Ocean. The response of the ocean is much more complex because the geometry of the basins has also changed.

Finally, the uplift of the landforms in Asia indirectly disrupts the climate by affecting the sources (oxidation) and sinks (chemical alteration and burial of organic matter) of carbon.

Can these mechanisms be transposed to other orogenies? The impact of orogeny on climate depends on the characteristics of the range (horizontal and vertical extensions) and the climatic context. Let us take the example of the Andes in South America. This range deflects the trade winds coming from the equatorial Atlantic Ocean into a powerful northerly flow (South American Low-Level Jet) transporting moisture from the Amazon basin to the subtropics and the Andean piedmont. This results in significant orographic precipitation in summer on the eastern flank of the Andes up to 30°S. Insel et al. (2010) shows with a regional climate model that reducing the Andean range elevation to half the present day induces a strong attenuation of the South American Low-Level Jet and suppression of moisture transport to the Andean foothills, resulting in an increase in precipitation over the Amazon basin. These results are in agreement with different paleoclimatic indicators (fauna, flora) that show an increase in precipitation around 10 Ma on the eastern flank of the Andes. The increase in precipitation is reflected by an acceleration of the exhumation rates inferred from fission tracks on apatite in the Central Andean thrust belt (Barnes et al. 2012). On the Pacific side, the semi-arid climate is replaced by a hyper-arid climate, ending supergene weathering (Strecker et al. 2007).

The impact of orogenies on climate has undoubtedly evolved since the earliest ages of the Earth as landforms have changed. In the late Archaean, high crustal and mantle radiogenic heat production as well as numerous magmatic intrusions within the early cratons limit the lithosphere's strength to support high topography (Rey and Houseman 2006). Rey and Coltice (2008) estimates that landforms should not have exceeded 1,800–2,000 m assuming a modern deformation rate (10^{-15}–10^{-14} s^{-1}). During the Proterozoic, secular cooling of the Earth favors a thicker and more rigid lithosphere favoring deformations at the boundaries of cratons. The orogenies at that time do not yet have the characteristics of modern orogenies, which could mean lower elevation than today. Ultra-high-pressure metamorphism (notably the blue shale facies) which is frequently found in the collision chains (see Volume 2 – Chapter 4) and is synonymous of modern-type orogenies, only dates back from the Neoproterozoic (approximately 600–800 Ma ago).

1.6. References

Acosta, R.P. and Huber, M. (2020). Competing topographic mechanisms for the summer Indo-Asian Monsoon. *Geophys Res Lett*, 47(3), e2019GL085112.

Barbolini, N., Woutersen, A., Dupont-Nivet, G., Silvestro, D., Tardif, D., Coster, P.M.C., Meijer, N., Chang, C., Zhang, H.X., Licht, A. et al. (2020). Cenozoic evolution of the steppe-desert biome in Central Asia. *Sci Adv*, 6(41), eabb8227.

Barnes, J.B., Ehlers, T.A., Insel, N., McQuarrie, N., Poulsen, C.J. (2012). Linking orography, climate, and exhumation across the central Andes. *Geology*, 40(12), 1135–1138.

Bhatia, H., Srivastava, G., Spicer, R.A., Farnsworth, A., Spicer, T.E.V., Mehrotra, R.C., Spicer, T.E.V., Bera, S., Valdes, P. (2021). Leaf physiognomy records the Miocene intensification of the South Asia Monsoon. *Glob Planet Change*, 196, 103365.

Bolin, B. (1950). On the influence of the Earth's orography on the general character of the Westerlies. *Tellus*, 2(3), 184–195.

Boos, W.R. and Kuang, Z. (2010). Dominant control of the South Asian monsoon by orographic insulation versus plateau heating. *Nature*, 463(7278), 218–222.

Botsyun, S., Sepulchre, P., Donnadieu, Y., Risi, C., Licht, A., Caves Rugenstein, J.K. (2019). Revised paleoaltimetry data show low Tibetan Plateau elevation during the Eocene. *Science*, 363(6430), eaaq1436.

Chamberlin, T.C. (1899). An attempt to frame a working hypothesis of the cause of glacial periods on an atmospheric basis. *J Geol*, 7(6), 545–584.

Champagnac, J.-D., Molnar, P., Sue, C., Herman, F. (2012). Tectonics, climate, and mountain topography. *J Geophys Res Solid Earth*, 117, B02403.

Charney, J. and Eliassen, A. (1949). A numerical method for predicting the perturbations of the middle latitude westerlies. *Tellus*, I(2), 38–54.

Clift, P.D. and Webb, A.A.G. (2019). History of the Asian monsoon and its interactions with solid earth tectonics in Cenozoic South Asia. *Geol Soc Lond Spec Publ*, 483, 875–880.

Dettman, D.L., Kohn, M.J., Quade, J., Ryerson, F.J., Ojha, T.P., Hamidullah, S. (2001). Seasonal stable isotope evidence for a strong Asian monsoon throughout the past 10.7 my (2001). *Geology*, 29, 31–34.

Ding, Y. and Krishnamurti, T. (1987). Heat budget of the Siberian High and the winter monsoon. *Mon Weather Rev*, 115, 2428–2449.

Ding, L., Spicer, R., Yang, J., Xu, Q., Cai, F., Li, S., Lai, Q.,Wang, H., Spicer, T., Yue, Y. et al. (2017). Quantifying the rise of the Himalaya orogen and implications for the South Asian monsoon. *Geology*, 45(3), 215–218.

Fang, X., Zan, J., Appel, E., Lu, Y., Song, C., Dai, S., Tuo, S. (2015). An Eocene–Miocene continuous rock magnetic record from the sediments in the Xining Basin, NW China: Indication for Cenozoic persistent drying driven by global cooling and Tibetan Plateau uplift. *Geophys J Int*, 201(1), 78–89.

Fang, X., Yan, M., Zhang, W., Nie, J., Han, W., Wu, F., Song, C., Zhang, T., Zan, J., Yang, Y. (2021). Paleogeography control of Indian monsoon intensification and expansion at 41 Ma. *Sci Bull*, 66(22), 2320–2328.

Farnsworth, A., Lunt, D.J., O'Brien, C.L., Foster, G.L., Inglis, G.N., Markwick, P., Pancost, R.D., Robinson, S.A. (2019). Climate sensitivity on geological timescales controlled by nonlinear feedbacks and ocean circulation. *Geophy Res Lett*, 46(16), 9880–9889.

Fluteau, F., Ramstein, G., Besse, J. (1999). Simulating the evolution of the Asian and African monsoons during the past 30Myr using an atmospheric general circulation model. *J Geophys Res Atmos*, 104(D10), 11995–12018.

France-Lanord, C. and Derry, L.A. (1997). Organic carbon burial forcing of the carbon cycle from Himalayan erosion. *Nature*, 390(6655), 65–67.

Gebelin, A., Mulch, A., Teyssier, C., Jessup, M.J., Law, R.D., Brunel, M. (2013). The Miocene elevation of Mount Everest. *Geology*, 41(7), 799–802.

Guo, Z.T., Sun, B., Zhang, Z.S., Peng, S.Z., Xiao, G.Q., Ge, J.Y., Hao, Q.Z., Qiao, Y.S., Liang, M.Y., Liu, J.F. et al. (2008). A major reorganization of Asian climate by the early Miocene. *Clim Past*, 4(3), 153–174.

Gupta, A.K., Yuvaraja, A., Prakasam, M., Clemens, S.C., Velu, A. (2015). Evolution of the South Asian monsoon wind system since the late Middle Miocene. *Palaeogeogr Palaeoclimatol Palaeoecol*, 438, 160–167.

Hahn, D.G. and Manabe, S. (1975). The role of mountains in the South Asian monsoon circulation. *J Atmos Sci*, 32, 1515–1541.

Halley, E. (1686). An historical account of the trade winds and monsoons observable in the seas between and near the tropics with an attempt to assign a physical cause of the said winds. *Philos Trans R Soc London*, 16, 153–168.

Hersbach, H., Bell, B., Berrisford, P., Biavati, G., Horanyi, A., Munõz Sabater, J., Nicolas, J., Peubey, C., Radu, R., Rozum, I. et al. (2019). ERA5 monthly averaged data on single levels from 1979 to present. Copernicus Climate Change Service (C3S) Climate Data Store (CDS). 10.24381/cds.f17050d7 [Online]. Available at: https://www.ecmwf.int/en/ forecasts/dataset/ecmwf-reanalysis-v5 [Accessed 15 March 2022].

Hilton, R.G. and West, A.J. (2020). Mountains, erosion and the carbon cycle. *Nat Rev Earth Environ*, 1(6), 284–299.

Hutchinson, D.K., Coxall, H.K., Lunt, D.J., Steinthorsdottir, M., de Boer, A.M., Baatsen, M., von der Heydt, A., Huber, M., Kennedy-Asser, A.T., Kunzmann, L. et al. (2021). The Eocene-Oligocene transition: A review of marine and terrestrial proxy data, models and model-data comparisons. *Clim Past*, 17, 269–315.

Huyghe, P., Bernet, M., Galy, A., Naylor, M., Cruz, J., Gyawali, B.R., Gemignani, L., Mugnier, J.L. (2020). Rapid exhumation since at least 13 Ma in the Himalaya recorded by detrital apatite fission-track dating of Bengal fan (IODP Expedition 354) and modern Himalayan river sediments. *Earth and Planet Sci Lett*, 534, 116078.

Insel, N., Poulsen, C.J., Ehlers, T.A. (2010). Influence of the Andes Mountains on South American moisture transport, convection, and precipitation. *Clim Dyn*, 35, 1477–1492.

Kutzbach, J.E., Guetter, P.L., Ruddiman, W.F., Prell, W.L. (1989). Sensitivity of climate to Late Cenozoic uplift in Southern Asia and the American West: Numerical experiments. *J Geophys Res*, 94, 18393–18407.

Ladant, J.-B., Donnadieu, Y., Dumas, C. (2014). Links between CO2, glaciation and water flow: Reconciling the Cenozoic history of the Antarctic Circumpolar Current. *Clim Past*, 10(6), 1957–1966.

Larsen, I.J., Montgomery, D.R., Greenberg, H.M. (2014). The contribution of mountains to global denudation. *Geology*, 42(6), 527–530.

Licht, A., van Cappelle, M., Abels, H.A., Ladant, J.-B., Trabucho-Alexandre, J., France-Lanord, C., Donnadieu, Y., Vandenberghe, J., Rigaudier, T., Lécuyer, C. et al. (2014). Asian monsoons in a late Eocene greenhouse world. *Nature*, 513(7519), 501–506.

Maffre, P., Ladant, J.B., Donnadieu, Y., Sepulchre, P., Goddéris, Y. (2018). The influence of orography on modern ocean circulation. *Clim. Dyn*, 50, 1277–1289.

Manabe, S. and Terpstra, T.B. (1974). The effects of mountains on the general circulation of the atmosphere as identified by numerical experiments, *J Atmos Sci*, 31, 3–42.

McGuffie, K. and Henderson-Sellers, A. (2005). *A Climate Modelling Primer*, 3rd edition. John Wiley & Sons, Chichester, West Sussex; Hoboken, NJ.

Meijer, N., Dupont-Nivet, G., Abels, H.A., Kaya, M.Y., Licht, A., Xiao, M.M., Zhang, Y., Roperch, P., Poujol, M., Lai, Z.P. et al. (2019). Central Asian moisture modulated by proto-Paratethys Sea incursions since the early Eocene. *Earth Planet Sci Lett*, 510, 73–84.

Mintz, Y. (1965). Very long-term global integration of the primitive equations of atmospheric motion. *WMO Tech Note*, 66, 141–167.

Mulch, A. (2016). Stable isotope paleoaltimetry and the evolution of landscapes and life. *Earth Planet Sci Lett*, 433, 180–191

Quade, J., Cerling, T.E., Bowman, J.R. (1989). Development of Asian monsoon revealed by marked ecological shift during the latest Miocene in Northern Pakistan. *Nature*, 342, 163–166.

Rey, P.F. and Coltice, N. (2008). Neoarchean strengthening of the lithosphere and the coupling of the Earth's geochemical reservoirs. *Geology*, 36, 635–638.

Rey, P.F. and Houseman, G. (2006). Lithospheric scale gravitational flow: The impact of body forces on orogenic processes from Archaean to Phanerozoic. *Geol Soc Lond Spec Pub*, 253(1), 153–167.

Rowley, D.B. and Garzione, C.N. (2007). Stable isotope-based paleoaltimetry. *Annu Rev Earth Planet Sci*, 35, 463–508.

Ruddiman, W.F. and Kutzbach, J.E. (1989). Forcing of late Cenozoic northern hemisphere climate by plateau uplift in southern Asia and the American west. *J Geophys Res*, 94, 18409–18427.

Sarr, A.C., Donnadieu, Y., Bolton, C.T., Ladant, J.B., Licht, A., Fluteau, F., Laugié, M., Tardif, D., Dupont-Nivet, G. (2022). Neogene South Asian monsoon rainfall and wind histories diverged due to topographic effects. *Nat Geosci*, 15, 314–319.

Sepulchre, P., Caubel, A., Ladant, J.-B., Bopp, L., Boucher, O., Braconnot, P., Brockmann, P., Cozic, A., Donnadieu, Y., Dufresne, J.-L. et al. (2020). IPSL-CM5A2 – An Earth system model designed for multi-millennial climate simulations. *Geosci Model Dev*, 13(7), 3011–3053.

Sorrel, P., Eymard, I., Leloup, P.-H., Maheo, G., Olivier, N., Sterb, M., Gourbet, L., Wang, G., Jing, W., Lu, H. et al. (2017). Wet tropical climate in SE Tibet during the Late Eocene. *Sci Rep*, 7(1), 1–11.

Spicer, R.A., Yang, J., Herman, A., Kodrul, T., Aleksandrova, G., Maslova, N., Spicer, T., Ding, L., Xu, Q., Shukla, A. et al. (2017). Paleogene monsoons across India and South China: Drivers of biotic change. *Gondwana Res*, 49, 350–363.

Spicer, R.A., Hoorn, C., Perrigo, A., Antonelli, A. (2018). Phytopaleoaltimetry: Using plant fossils to measure past land surface elevation. In *Mountains, Climate and Biodiversity*, Hoorn, C., Perrigo, A., Antonelli, A. (ed.). Wiley, Oxford.

Strecker, M.R., Alonso, R.N., Bookhagen, B., Carrapa, B., Hilley, G.E., Sobel, E.R., Trauth, M.H. (2007). Tectonics and climate of the southern Central Andes. *Ann Rev Earth Planet Sci*, 35, 747–787.

Su, T., Spicer, R.A., Wu, F.X., Farnsworth, A., Huang, J., Rio, C.D., Deng, T., Ding, L., Deng, W.Y.D., Huang, Y.J. et al. (2020). A Middle Eocene lowland humid subtropical "Shangri-La" ecosystem in central Tibet. *PNAS*, 117(52), 32989–32995.

Sun, X. and Wang, P. (2005). How old is the Asian monsoon system? Palaeobotanical records from China. *Palaeogeogr Palaeoclimatol Palaeoecol*, 222(3–4), 181–222.

Tardif, D., Fluteau, F., Donnadieu, Y., Le Hir, G., Ladant, J.-B., Sepulchre, P., Licht, A., Poblete, F., Dupont-Nivet, G. (2020). The origin of Asian monsoons: A modelling perspective. *Clim Past*, 16(3), 847–865.

Tauxe, L. and Feakins, S.J. (2020). A reassessment of the chronostratigraphy of Late Miocene C3–C4 transitions. *Paleoceanogr Paleoclimatol*, 35, e2020PA003857.

Webster, P.J., Magana, V.O., Palmer, T.N., Shukla, J., Tomas, R.A., Yanai, M., Yasuna, T. (1998). Monsoons: Processes, predictability, and the prospects for prediction. *J Geophys Res*, 103(C7), 14451–14510.

Westerhold, T., Marwan, N., Drury, A.J., Liebrand, D., Agnini, C., Anagnostou, E., Barnet, J.S.K., Bohaty, S.M., De Vleeschouwer, D., Florindo, F. et al. (2020). An astronomically dated record of Earth's climate and its predictability over the last 66 million years. *Science*, 369(6509), 1383–1387.

Yang, H. and Wen, Q. (2020). Investigating the role of the Tibetan plateau in the formation of Atlantic Meridional Overturning Circulation. *J Clim*, 33(9), 3585–3601.

Zhao, L., Lu, H.Y., Wang, H.L., Meadows, M., Ma, C.M., Tang, L.Y., Lei, F., Zhang, H.Y. (2020). Vegetation dynamics in response to evolution of the Asian Monsoon in a warm world: Pollen evidence from the Weihe Basin, central China. *Glob Planet Change*, 193, 103269.

Zoura, D., Hill, D., Dolan, A., Hunter, S., Tang, Z., Haywood, A. (2019). Atmospheric carbon dioxide, ice sheet and topographic constraints on palaeo moisture availability in Asia. *Earth Planet Sci Lett*, 519, 12–27.

2

Eroding the Himalaya: Processes, Evolution, Implications

Vincent GODARD[1], Mikaël ATTAL[2], Saptarshi DEY[3], Maarten LUPKER[4] and Rasmus C. THIEDE[5]

[1] *CEREGE, Aix-en-Provence, France*

[2] *University of Edinburgh, United Kingdom*

[3] *Indian Institute of Technology, Gandhinagar, India*

[4] *Independent researcher, Bern, Switzerland*

[5] *Christian-Albrecht University, Kiel, Germany*

2.1. Introduction

The action and impact of denudation on mountain landscapes have been a central question in orogenic process research over the last few decades. Processes acting at the Earth surface, such as erosion, weathering or sediment transport, contribute to a fundamental coupling between internal and external geodynamics, with far-reaching implications for tectonics and global climate (see Volume 3 – Chapter 1). The Himalaya play a prominent role in this debate as its erosion produces a large fraction of the global sediment delivered to neighboring ocean basins.

Himalaya, Dynamics of a Giant 3,
coordinated by Rodolphe CATTIN and Jean-Luc EPARD.

The Himalaya have also been one of the key natural laboratories for the study of the interactions between erosion, climate and tectonics due to a number of first-order features which make this range one of the most singular geological settings on Earth. As one of the largest and most active orogenic systems, the Himalaya display relief and topographic gradients which are unmatched anywhere else, and in turn control very intense surface processes such as river incision or mass wasting. Furthermore, the Himalayan climate is dominated by a long-lasting powerful monsoon regime, bringing intense precipitation highly focused on its southern flank, which promotes hillslope failure and weathering. Himalayan rivers, which drive the erosion of the range and export a massive amount of sediment, are also supporting one of the most populated areas of the world in the Indo-Gangetic plains.

All these characteristics lend themselves to asking more in-depth questions about the processes that control erosion at the scale of an entire orogen. Notably, what sets the distribution of denudation rates across the Himalaya? What are the respective contributions of extreme tectonic and climatic events? And how is erosion intensity modulated by long-term changes in tectonic and climatic forcings? This chapter makes an attempt at highlighting the most notable characteristics of Himalayan erosion, such as the main processes involved in and the spatial patterns of erosion, the coupling of erosion with tectonic and climatic forcing, as well as the temporal variability of these processes over periods ranging from single events to millions of years.

2.2. Main process domains in the Himalaya

We present here the three families of surface processes responsible for eroding the Himalayan range. They operate on different spatial and temporal scales, and their relative contribution to the production and transport of sediment varies across the range.

2.2.1. *Himalayan rivers*

Like all mountainous landscapes, the Himalayan range is dissected by river networks which are the main drivers of denudation and landscape evolution. Rivers incise into bedrock and cut deep valleys: they set the pace of erosion for the adjacent hillslopes and transport the sediment produced by various processes out of the range. From North to South, a typical Himalayan river will incise into the steep high-relief and high-uplift domains of the High Range, then cross the lower uplift area of the Lesser Himalaya, before reaching the weak sedimentary series of the Siwalik foothills above the Main Frontal Thrust (Figures 2.1 and 2.2). While many Himalayan rivers have their headwaters in the High Range, the largest trans-Himalayan rivers are actually connected to Southern Tibet and can have significant parts of their drainage on the low relief Tibetan Plateau.

STREAM POWER.– To erode rocks and transport sediment, rivers need power. "Stream power" is a function of the slope, discharge and width of a river: a steep, narrow river with a high discharge will have more power than a low gradient, wide river with a low discharge (Figure 2.3). A rule of thumb is that rivers with the greatest power tend to be those that erode fastest, although the relationship between erosion rate and stream power is modulated by bedrock resistance (a river will need more power to erode into harder rocks) and the availability of sediment grains (which constitute the "tools" for erosion).

All Himalayan rivers are connected to the base level of the Indo-Gangetic plains which can be considered as a stable reference frame above which the relief of the range is uplifted. One of the key aspects of Himalayan river dynamics is that, over >1 ka timescales, their incision rate into bedrock is highly dependent on the distribution of rock uplift across the range (Lavé and Avouac 2001). This adjustment of incision rates to rock uplift is reflected into the rivers' long-profile morphology, with notably much steeper reaches where rivers cross the High Range (Figure 2.3). Such changes in river long-profile steepness have long been recognized as a very useful proxy to detect changes in patterns of tectonic activity, and Himalayan rivers have been used intensively to develop this methodology (e.g. Seeber and Gornitz 1983).

One of the main features of the climatic system of the Himalaya is the monsoon regime, which causes a very contrasted distribution of precipitation throughout the year, with ca. 80% of rainfall occurring between May and September (Bookhagen and Burbank 2010). This uneven distribution of precipitation during the year directly controls river discharge, with much more intense flow during the summer, which is when most incision and sediment transport occurs (Scherler et al. 2017). The high topographic barrier of the Himalaya also has a direct impact on the distribution of precipitation, with an important orographic effect associated with the High Range (Figures 2.2 and 2.4).

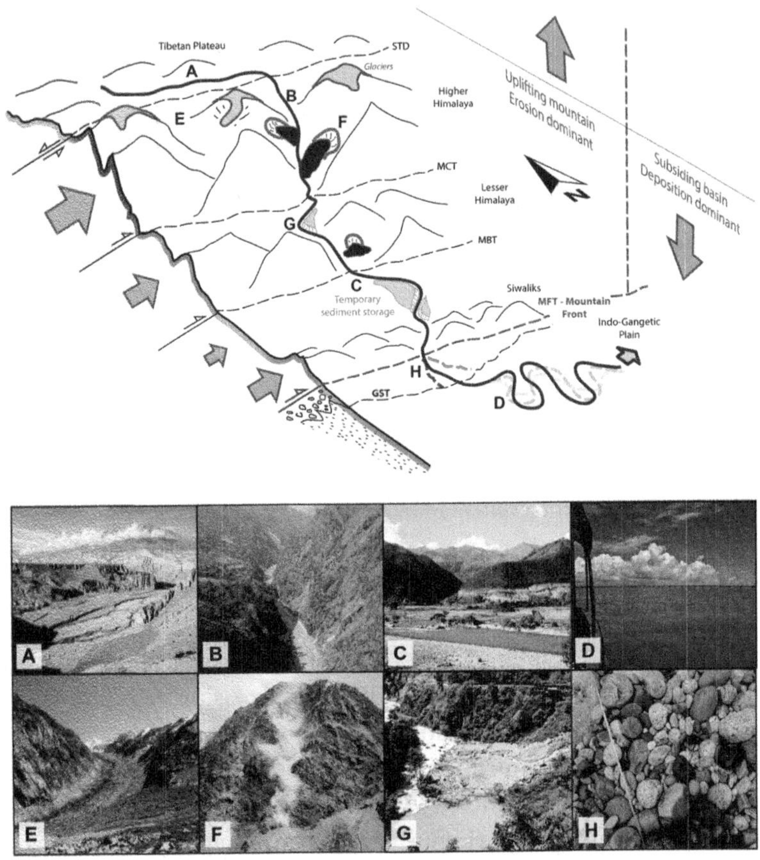

Figure 2.1. *Idealized trans-Himalayan river system, illustrating the different morphological domains. See Volume 1 – Chapter 3 for the meaning of STD, MCT, MBT and MFT. GST is the gravel–sand transition. Thick dashed lines downstream of the MFT represent paleo-channels. Colors along the topographic profile to the left express local steepness, from green (low gradient) to red (steep gradient). Letters A–H refer to photos in the lower panel. A – Rivers in the Tibetan headwaters exhibit low gradients and store large amounts of sediments. B – Upon entering the Higher Himalaya, river channels steepen and local relief increases. C – In the Lesser Himalaya, channel slopes become more gentle with the reappearance of intermittent sediment storage. D – Beyond the MFT, rivers enter the alluvial domain with very low gradients. E – At elevations above ca. 5,000 m, glacial and peri-glacial processes contribute significantly to erosion processes. F – Steep slopes throughout the range are subject to frequent landsliding during monsoons or earthquakes. G – Eroded sediments are efficiently transferred to the fluvial network and are exported out of the range. H – The sediment river sediment load integrates upstream lithologies and surface processes. For a color version of this figure, see www.iste.co.uk/cattin/himalaya3.zip*

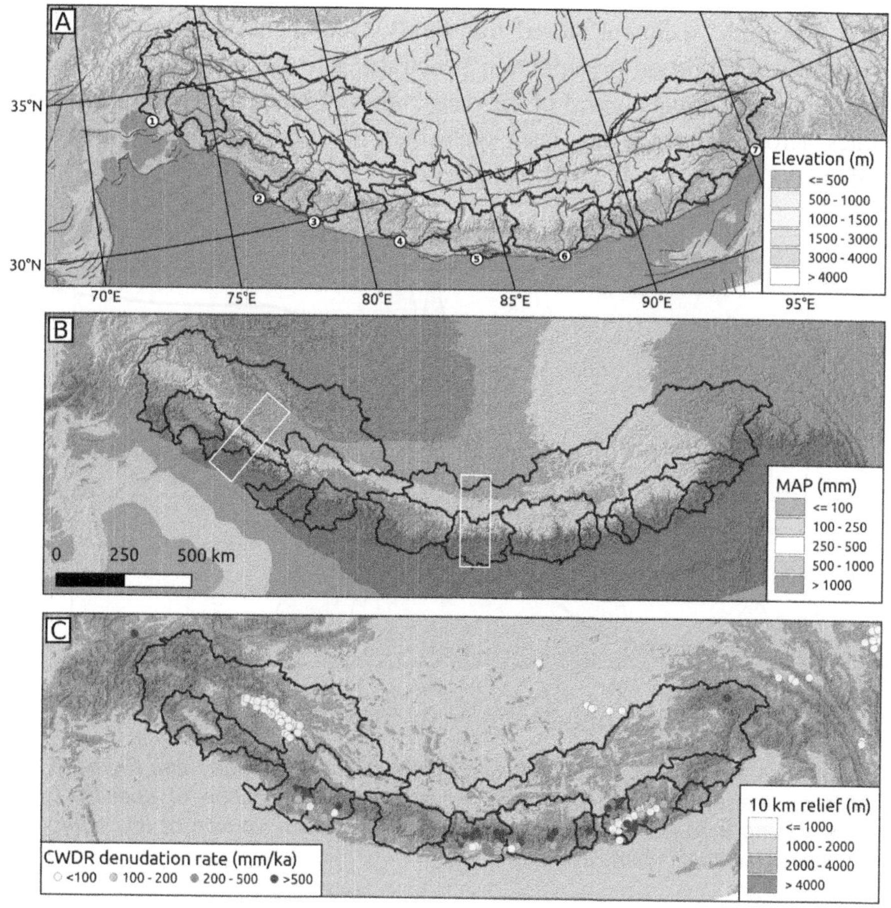

Figure 2.2. *Maps of the Himalayan arc. (A) Topographic map of the Himalaya with main faults (red), rivers (blue) and major catchment outlines (black). 1: Indus, 2: Sutlej, 3: Ganges, 4: Karnali, 5: Narayani, 6: Sapt Kosi, 7: Brahmaputra. (B) Mean annual precipitation (Fick and Hijmans 2017). White rectangles show the location of swath profiles in Figure 2.4. (C) Relief computed over a 10 km window. Circles are catchment-wide denudation rates (CWDR) derived from* ^{10}Be *concentrations in river sediments from the Octopus database (Codilean et al. 2018), for basins with areas in the 20–300 km*^{2} *range. For a color version of this figure, see www.iste.co.uk/cattin/himalaya3.zip*

The Himalaya have one of the most efficient fluvial systems on Earth in terms of incision and sediment transport (with Taiwan and the Southern Alps of New Zealand). It displays some of the fastest incision rates recorded, of the order of 10 mm/yr, for example, in the Siwaliks (Lavé and Avouac 2000), and very little sediment stored transiently around the southern Himalayan front where monsoon precipitation penetrates into the orogen. However, the existence of a direct control of the distribution of precipitation on denudation rates across the range has been questioned in various locations across the Himalaya (Godard et al. 2014; Scherler et al. 2014).

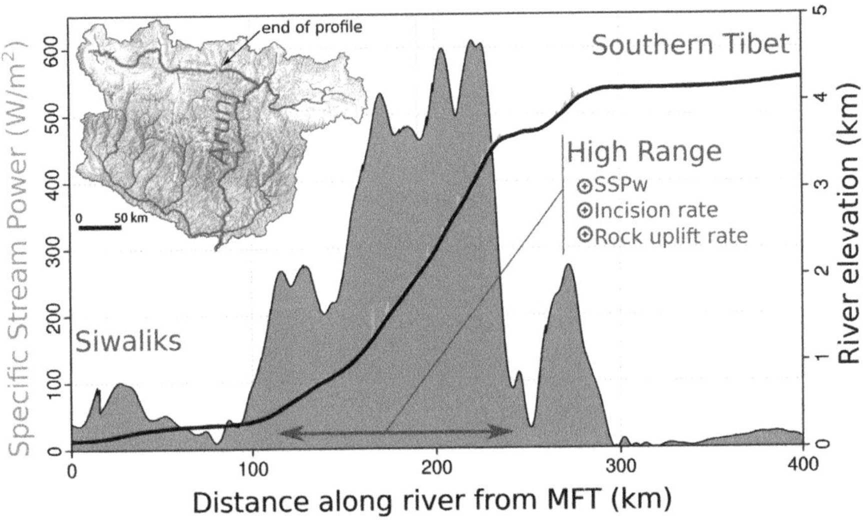

Figure 2.3. *Arun River catchment and profile (Sapt Kosi basin; see Figure 2.2(A)) showing the evolution of river bed elevation and distribution of specific stream power along a typical trans-Himalayan river. For a color version of this figure, see www.iste.co.uk/cattin/himalaya3.zip*

Himalayan rivers are dominantly detachment-limited systems, where the main control on channel evolution is the capacity of the river to detach particles of rock from the bedrock. Incision occurs due to processes such as abrasion or plucking driven by repeated impacts of moving bedload upon the river bottom. Himalayan rivers are also the main agents ensuring the transport of sediment across and out of the range.

The occurrence of massive fill terraces, notably in the Lesser Himalaya, indicates strong fluctuations in both sediment supply from upstream and the capacity of Himalayan rivers (which is related to stream power; see box) to accommodate and transfer this load downstream (Figure 2.1).

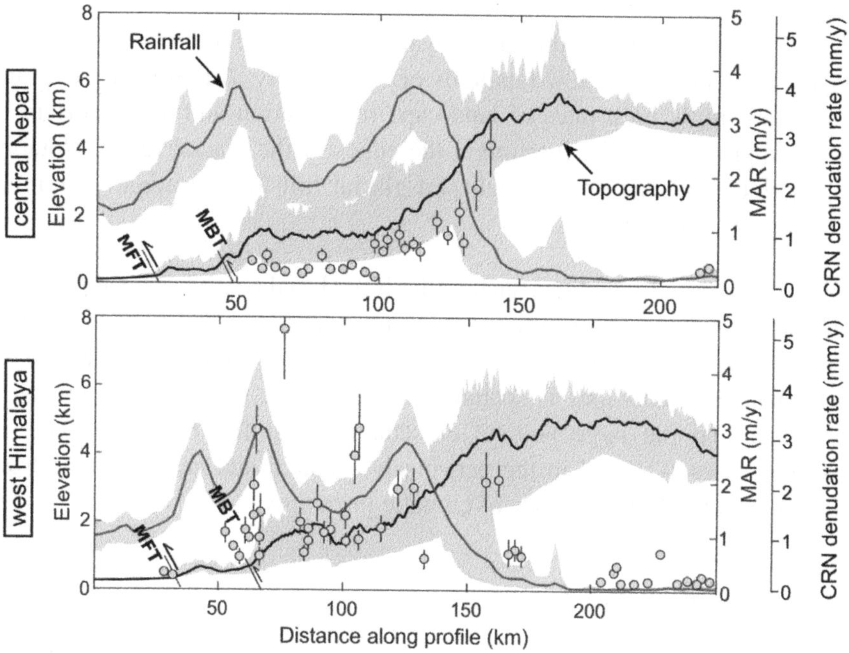

Figure 2.4. *Swath profiles showing variations in topography, rainfall and present-day catchment-wide denudation rates across the central and the western Himalaya (see Figure 2.2 for location). Present-day catchment-wide denudation rates are derived from CRN (^{10}Be) concentrations in river sands (see box "Cosmogenic nuclides and denudation"). Note the common locations of the orographic barrier, high topographic relief and hotspot for denudation. For a color version of this figure, see www.iste.co.uk/cattin/himalaya3.zip*

2.2.2. *The glaciated High Range*

With an average elevation >5,000 m in the High Range, the Himalaya host an extensive glacial domain with well-developed valley glacier systems (see Volume 3 – Chapter 3). Glaciers are powerful agents of erosion, sediment production and modification of Himalayan landscapes at high elevation, even if the actual amplitude of glacial erosion is still poorly constrained in the Himalaya. Widespread occurrence of U-shaped valleys testifies for the advance of glaciers in full glacial conditions and their morphological impact, notably for maintaining high relief in some parts of the range. However, the low-latitude position of the range did not allow the extension of glacier systems to low-elevation areas (Owen and Dortch 2014). There are actually few reliable estimates of the depression of the Equilibrium Line Altitude during the Last Glacial Maximum along the Himalayan arc, and the limited observations suggest a lowering of a few hundreds of meters only. Additionally, the high uplift rates imply high rates of landscape rejuvenation and quick erasure of glacial imprint by fluvial incision at the edges of the glacial realm. Notably, the imprint of glaciations through peri-glacial hillslope weathering is more limited than in other ranges.

In some parts of the range, such as the NW Himalaya, it has been proposed that glacial processes were imposing a strong limitation on the development of topography (Brozovic et al. 1997). The direct imprint of glacial processes on Himalayan landscapes is spatially limited to the high elevation parts of the range, and the large volume of sediment that has been produced by these processes is stored in the peri-glacial domain and can be remobilized catastrophically, during periods of climatic instability.

2.2.3. *Critical hillslopes*

While most long-distance sediment transport is done by rivers, sediment is dominantly produced on hillslopes adjacent to the river network. Like in most high-relief mountain ranges, weathering processes transform and weaken bedrock and hillslope gravitational processes transport the sediments downslope. As a mountain range spanning a large range of elevations and located at mid-latitudes, the Himalaya experience a wide variety of climatic conditions, from the hot and humid tropical Siwalik foothills to the cold and dry Alpine environments of the High Range.

This diversity of conditions implies a spectrum of hillslope regimes in terms of the intensity of weathering and the nature of transport processes. However, as the Himalaya are one of the most rapidly uplifting mountain ranges in the world, hillslope angles are high and often close to failure thresholds throughout the range (Gabet et al. 2004b), which results in high sediment transport rates across hillslopes, notably by catastrophic mass movements such as landslides or debris flows (Larsen and Montgomery 2012). The occurrence of these intense events can be clustered in time in relation to short-term climatic and tectonic forcings such as earthquakes and extreme monsoon rainfall but they also occur stochastically.

Over timescales >10 ka, landscape evolution is highly dependent on the coupling between river incision and hillslope denudation. For example, rapid incision in the High Range induces the formation of deep valleys, with steepened hillslopes promoting faster mass transport processes and landsliding. The sediments delivered by the hillslopes are then transported by the river system, and their volume and grain size distribution will have a first-order control on fluvial processes such as sediment transport and incision into bedrock. The feedback between hillslopes and rivers is one of the most important elements of landscape dynamics, and is particularly strong in the Himalaya due to the very high rates of river incision and hillslope erosion (Burbank et al. 1996). The grain size of the sediment delivered by hillslope processes varies widely across the range. The material supplied by landslides in the steep and rapidly uplifting High Range tends to be coarsest, with most rivers in the High Range containing sediment that commonly includes megablocks several meters wide (Attal and Lavé 2006).

2.3. Extreme events and their contribution to denudation

Denudation processes slowly shape landscapes over millions of years, but this long-term evolution results in many cases from the incremental contribution of brief and intense events. One key area of research in tectonically active landscapes aims to better constrain the role of short-duration high-intensity extreme events, such as earthquakes, storms or lake outburst floods, in the global erosion budget and the shaping of mountain ranges.

Concerning earthquakes, it is often proposed that a significant fraction of long-term erosion is temporally clustered during and after seismic shaking events which destabilize hillslopes. One of the best examples of such behavior is the 2008 $M_w \sim 7.9$ Wenchuan earthquake in Eastern Tibet, which triggered massive landsliding throughout the area affected by co-seismic shaking, with an overall volume of eroded material estimated by some studies to be of the same order of magnitude as the uplifted rock volume during the earthquake (Li et al. 2014).

In the central Himalaya, the effects of the recent $M_w \sim 7.8$ Gorkha (April 25, 2015) earthquake are still being intensively studied (see Volume 3 – Chapter 8). The Gorkha earthquake ruptured part of the Main Himalayan

Thrust (MHT) in Central Nepal, directly north of Kathmandu and at the foot of the High Range. The co-seismic shaking triggered >25,000 co-seismic landslides representing a volume of sediments equivalent to approximately four to six years worth of inter-seismic erosion and with a spatial distribution closely related to that of the inferred seismic rupture (Roback et al. 2017). Surprisingly, this event has not led to significant increases in sediment fluxes from major drainage basins in the central Himalaya (Märki et al. 2021), further highlighting the role of hillslopes as transient sediment stores. Modeling studies integrating co-seismic landsliding over the entire Himalayan seismic cycle suggest that sediment production and erosion are dominated by medium-sized earthquakes (ca. $M_w 6.8$) (Marc et al. 2019).

POWER LAW DISTRIBUTION.– A common characteristics among episodic natural processes is that they often display a power law relationship between their magnitude M and their frequency or return time, f, such as $M \propto f^{-\alpha}$. One of the most famous examples in Earth Sciences is the magnitude–frequency relationship for earthquakes, known as the Gutenberg–Richter law. In geomorphology, landslide volume and river discharge are also good examples of quantities displaying such behavior. The characteristics of the power law, and in particular its exponent α, implies that large events may have disproportionately large impacts on the evolution of the landscape despite their rare occurrences.

Fluvial incision is another example of a process which builds incrementally over time by the summation of discrete intense events. The monsoon regime of the Himalaya implies that incision and sediment transport are focused during high discharge events restricted in time. While river channels are mostly adjusted to usual monsoon discharges, unusually intense monsoon precipitation may still disproportionately affect long-term erosion rates and patterns if considered over hundreds of thousand years (Deal et al. 2017). It is worth noting that the coupling between hillslopes and rivers is heightened during the monsoon, as intense monsoon rainfall events can also trigger catastrophic mass transport on hillslopes. The increase in pore pressure in the subsurface and runoff induces sudden slope failure and debris flows moving large volumes of bedrock and regolith down into the river system (Gabet et al. 2004a; Struck et al. 2015).

Probably, the most extreme examples of high discharge events are lake outburst floods from either the catastrophic failure of landslide dams or glacier-lake outburst floods (GLOFs; see Volume 3 – Chapter 3). About one to two significant GLOFs per year are currently observed across the Himalayan arc on average in modern times (Veh et al. 2019). The discharge generated by outburst floods significantly exceeds monsoon discharges in the upper reaches of the Himalayan watersheds, causing the mobilization of the largest grain sizes available on the river bed (Cook et al. 2018; Huber et al. 2020) and thereby promoting channel incision. Sediment deposits recording large GLOF events in the Eastern Himalayan syntaxis suggest that large discharge events can account for 1–4 ka worth of monsoonal-driven incision (Lang et al. 2013). A recent analysis of river profiles in central Nepal shows that the imprint of extreme GLOF events is pervasive (Dahlquist and West 2022), highlighting their key role in shaping Himalayan landscapes.

Evidences for widespread slope destabilization in response to extreme climatic events or earthquakes are numerous along the Himalayan arc, but there is also evidence for such responses in the past. For example, debris flow terraces are widespread along the rivers in the Lesser Himalaya, resulting from the extensive mobilization of hillslope material during intense rainfall in the High Range (Lavé and Avouac 2001). In central Nepal, massive deposits of sediment sourced from the High Range have been identified in the Lesser Himalaya near Pokhara (Schwanghart et al. 2015). The sedimentology and geochronological constraints on these deposits suggest a catastrophic valley infill origin as a possible consequence of three medieval earthquakes. Evidence for very large landslides during the Holocene has been found at various locations in central Nepal, mobilizing volumes exceeding 1 km^3 (Marc et al. 2019).

2.4. 1–10 ka timescale and climatic oscillations

Pleistocene climate is characterized by wide climatic oscillations at 10–100 ka timescales, between glacial and interglacial conditions, which are controlled by the Earth's orbit parameters (Figure 2.5). The last ~800 ka have, for example, been dominated by cycles at 100 ka period mainly controlled by the eccentricity of the orbit. The associated changes in temperature and precipitation have major implications for every denudation processes acting in the Himalaya, from advancing and retreating glaciers to changes in the intensity and distribution of monsoon floods. One important characteristic of the Himalaya is the rapid rate of uplift/erosion which implies fast response times and rejuvenation for landscapes, such that geomorphologic climatic records (terraces, moraines, etc.) are usually not preserved inside the range beyond a few 10s ka (unlike the sedimentary archives in flexural basins and offshore that can keep records of such signals for millions of years). It is not expected that climatic fluctuations over such timescales can have a significant impact on channel long-profiles, so that most of the climatic impact on erosion over these timescales should arise from the direct modulation of sediment production and transport over hillslopes (Goren 2016; Godard and Tucker 2021).

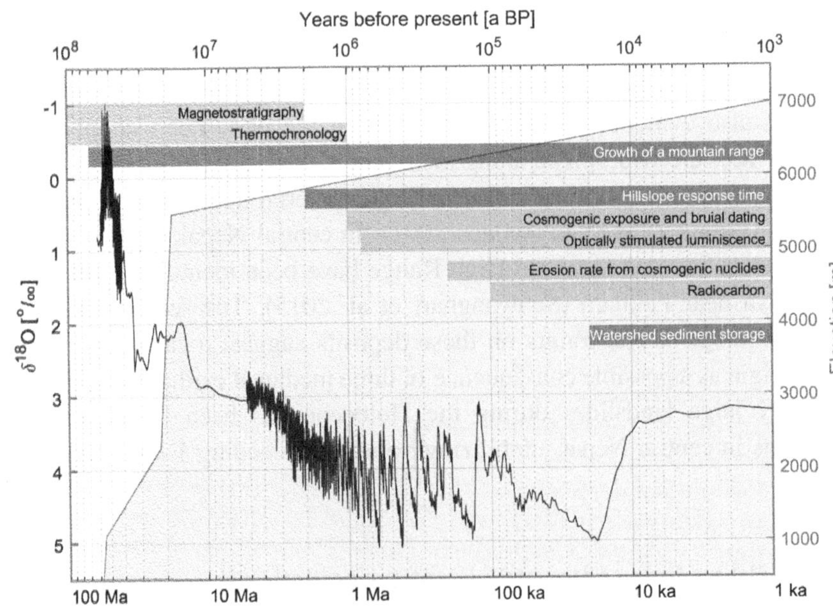

Figure 2.5. *Relative timescales of geological processes (gray) such as growth of mountain range, climate change and landscape evolution, and selected analytic methods (orange) studying these processes. The blue line represents a marine oxygen isotope record, a proxy for global temperature and ice volume, extending from approximately 70 Ma to 1 ka. The left edges of the green-shaded bar indicate approximate time spans of the uplift of the Himalayan orogen (Ding et al. 2017). Oxygen isotope data are from benthic stack compilations by Zachos et al. (2008) prior to 5 Ma and Lisiecki and Raymo (2005) between 5 Ma and the present. This figure is modified and redrafted after Perron (2017). For a color version of this figure, see www.iste.co.uk/cattin/himalaya3.zip*

The monsoonal regime is considered to be a long-standing feature of the Himalayan climatic system for at least the last 10 Ma (Dettman et al. 2001), implying a high seasonality of hillslope and fluvial denudation and sediment transport processes. Over the Holocene, episodes of strengthening and weakening of the Indian Summer Monsoon induced a corresponding modulation of sediment fluxes, which can, for example, be identified in the terrace record by phases of aggradation and incision (Bookhagen et al. 2005, 2006). Monsoon cycles in the Himalaya have a strong impact on the temporal variations in erosion rates. During strong monsoon phases, such as during the transition from MIS 2 to MIS 1 (Marine Isotope Stages) and MIS 6 to MIS 5 (Lisiecki and Raymo 2005), the average erosion rates could be as high as 3-4 mm/yr, while during weak monsoon phases or even at present-day monsoon conditions, erosion rates are usually < 1 mm/yr (e.g. Dey et al. 2022).

Interestingly, the range of erosion rates between strong and weak monsoon phases brackets the long-term (million-year timescale) exhumation rates (e.g. Thiede et al. 2017). High erosion rates during stronger monsoon underline the impact of short-lived extreme events on erosion and sediment transfer from the mountains to the basins.

Over thousands of years timescales, denudation rates have been derived in the Himalaya using cosmogenic nuclide concentrations (^{10}Be) in river sediments: they provide an estimate of the average denudation rate in the catchment upstream of the sampling point (Figure 2.2(C)). Reported denudation rates are highly variable along and across the range, from very low rates in the dry high elevation Ladakh range up to rates >1 mm/yr in some catchments of the High Himalaya. However, a clear pattern has been identified across the range, with a northward increase from the Lesser to the Higher Himalaya, which appears to be strongly correlated with the distribution of tectonic rock uplift (Figure 2.4). In the Garwhal Himalaya and Western Bhutan, the distribution of catchment-wide denudation rates has been used to make inferences on the geometry of the MHT (Scherler et al. 2014; Le Roux-Mallouf et al. 2015). Catchment-wide denudation rates do not appear to be significantly controlled by precipitation over ka timescales (Godard et al. 2014; Scherler et al. 2014).

COSMOGENIC NUCLIDES AND DENUDATION.– In order to measure landscape denudation rates in mountain ranges, the concentration of rare atoms called cosmogenic radionuclides (CRN) is measured in sediment. Cosmic rays constantly bombard the Earth and when they hit atoms inside rocks exposed at the surface of the Earth, they can create very small amounts of new atoms which would otherwise not exist, such as, for example, ^{10}Be. Rocks start accumulating CRN within their minerals when they are close to or at the surface of the Earth (a few meters). CRN abundance can be used like a stopwatch: the longer a rock sits near the surface of the Earth, the greater the CRN concentration in its minerals. If erosion rates are low, rocks will sit at the surface of the Earth for a long time before they get turned into sand, so river sand (which comes from the erosion of rocks) will contain a lot of CRN. Conversely, if the erosion rates are high, rocks will be eroded away rapidly and will not have the time to accumulate much of these atoms: CRN concentration in river sand will be low.

At the scale of a few millennia, the landscape of the lower part of the Himalaya has been intensely modified by human activities and the development of agriculture, including deforestation, terracing and rice paddy development. However, comparison of present-day sediment fluxes with background longer-term denudation rates from CRN measurements in river sediments (integrating over several millennia) does not show a clear increase which could be associated with anthropogenic disturbances of the landscape (West et al. 2015).

2.5. Impact of long-term tectonic and climatic evolution over several Ma

Constraining the amount of denudation across a mountain range on million-year-timescales is challenging. Indeed, despite relatively steady rates of plate convergence, rock uplift rates typically vary in space and time, owing to the formation of discrete thrust faults that remain active for a finite amount of time (Konstantinovskaia and Malavieille 2005; Naylor and Sinclair 2007). Because landscapes need time to adjust to changes in rock uplift (Whipple and Meade 2006), tectonic perturbations ought to entail periodic changes in surface uplift and thus topography and erosion rates. The Himalaya-wide distribution of tectonic rock uplift as well as erosion are at first order imposed by the geometry of the MHT and related fault systems.The analysis of the thermal cooling history of rocks as a function of their burial and exhumation history is a key approach to reconstruct the long-term denudation history of orogenic belts (see Volume 1 – Chapter 6). A dense dataset for cooling ages, derived from low-temperature thermochronology, has enabled us to document strong spatial and temporal variations in mountain building, uplift and denudation across the range since the collision started ~50 Ma ago. Combining these datasets with thermal models of the range provided further information on the thrust faulting geometry and its relation to topography or denudation along various segments of the Himalaya (Vannay et al. 2004; Bollinger et al. 2006; Herman et al. 2010; Robert et al. 2011; Thiede and Ehlers 2013; Ghoshal et al. 2020). These results indicate a structural segmentation of the Himalayan front that is strongly influencing along-strike variations in denudation (see also Volume 1 – Chapter 6 for further reading).

While climatic fluctuations operate on shorter timescales, they can temporarily modulate the amplitude of erosion or incision processes, in particular in landscapes that are not far from dynamic equilibrium (when uplift rates match denudation rates). But the denudation rate is ultimately pegged to the rate at which the crustal material is uplifted above the range's base level, constituted by the Indo-Gangetic Plain in the Himalaya. For example, the exhumation and denudation patterns in central Nepal indicate a clear increase in surface process intensity when moving north from the Lesser Himalaya into the High Range, which is consistent with the flat-ramp geometry of the MHT and its northward steepening (Robert et al. 2009; Godard et al. 2014). The Indian summer monsoon is the first-order atmospheric circulation system controlling precipitation patterns across the Indian peninsula as well as the southern margin of the Tibetan plateau. These continental atmospheric circulation systems have been established prior to and strengthened throughout the early Miocene. Recent studies suggest that monsoonal precipitation and erosion were strongest during the middle Miocene, approximately correlating with the global middle Miocene climate optimum, and fostered the rapid exhumation along the southern flanks of the Greater or Higher Himalaya since then. The proposed weakening of monsoonal strength, a general regional drying and decreasing overall erosion during the Late Miocene are linked to general global cooling trends (Clift et al. 2008). While different proxy records indicate that monsoonal strength, precipitation and erosion went through a minimum during the Pliocene, they appear to have increased again during the Pleistocene.

LOW-TEMPERATURE THERMOCHRONOLOGY.– Low-temperature thermochronology is a group of rock dating techniques using the natural decay of isotopes such as ^{238}U, ^{235}U or ^{40}K and the changes in diffusion behavior of daughter isotopes through a crystal lattice with temperature, usually in the 40–350°C range (which corresponds to depths of a few kilometers below the surface). These methods are most commonly applied to the minerals apatite, zircon and mica, and measure the elapsed time since the transition from the total diffusional loss of daughter elements to the quantitative retention of these elements, which is referred to as a rock cooling age. Practically, these ages (usually in million years) can be understood as the time when the analyzed rocks moved through a certain temperature on their way towards the surface, and allow quantification of exhumation rates.

It is suspected that the pronounced evolution of the denudation patterns and topography observed along strike are direct consequences of modifications in the tectonic boundary conditions, notably in terms of MHT geometry and crustal structures (Bhutan, Western Nepal). Global analysis of thermochronological data in the Himalaya reveals exhumation patterns which are highly variable in space and time over the last few Ma and reflects the combined influence of the progressive tectonic evolution and climatically controlled focused erosion in the High Range (Thiede and Ehlers 2013; van der Beek et al. 2016). Recent studies further advance this idea by invoking the punctuated nature of tectonic accretion on ~1-Ma timescales to explain the measured temporal variability in erosion rates within the Himalaya (Mandal et al. 2021). These studies relate the inferred quasi-cyclic pattern in erosion rates to sequences of tectonic growth of the Himalaya, rather than to changes in climate.

Sedimentary series deposited at the range front in the foreland basins have also been used as records for the long-term evolution of the range (see Volume 2 – Chapter 8). Notably, the Siwalik detrital sediments, which are now deformed and incorporated into the range since the activation of the Main Frontal Thrust (MFT), provide information on the characteristics of denudation over the last ~12–16 Ma. These sedimentary archives close to the source catchments hold the potential to record much of the tectonically and climatically related erosion rate variability of the hinterland. For example, changes in the sources of sediments associated with drainage basin reorganization or to changes in the intensity of weathering linked to climatic evolution can be tracked through stratigraphy (Chirouze et al. 2015; Vögeli et al. 2018). More distant sedimentary records such as those preserved in the deep-sea fans also provide critical information on the evolution of orogen-wide denudation in the Himalaya. Interestingly, recent results for paleo-denudation obtained in the Bengal fan with CRN data (^{10}Be) indicate that the average denudation of the range was relatively stable through the last 6 Ma, suggesting a limited impact of the Late Cenozoic cooling (Lenard et al. 2020).

2.6. Out of the Himalaya: sediment transport and storage from the range to the sedimentary basin

Sediment is generated by erosion in mountains and evacuated by rivers to the sedimentary basins (Figure 2.6). Sediment accumulates and is transiently stored on the Indo-Gangetic Plain, where it influences the livelihood of around a billion people. Part of the sand fraction and finer particles of the sediment load will make it to the coast and contribute to the building of the Indus and Bengal fans. But where do the different sediment fractions (gravel, sand and silt) come from, and how far do they travel? In the following, we will explore sediment dynamics from the perspective of an ideal trans-Himalayan river (Figure 2.1).

On the Tibetan Plateau, rivers are not very powerful: their gradient is low and the lack of precipitation means that discharge is low too (Figure 2.4). At such high elevation, there is a strong glacial overprint, so rivers are mostly tapping into the glacial sediment. Across the Higher Himalaya, rivers are the most powerful: rivers are steep and narrow and precipitation is also focused on the Higher Himalaya. Rivers are tightly coupled to the hillslopes which ensures direct supply of sediment from hillslopes (via gravitational mass movements such as landsliding and debris-flows) and frequent transport. The Higher Himalaya constitutes the first-order source of sediment to the sedimentary basin. As rivers reach the Lesser Himalaya and Siwaliks, valleys widen and slope decreases, lowering the power of the rivers. These areas are still being actively uplifted, so erosion will prevail on the long-term. However, sediment can remain stored in terraces or intermontane basins ("duns") over timescales of the order of 1–100 ka (Blöthe and Korup 2013). Note that the MFT is accommodating a large part of the India–Asia convergence (~20 mm/yr). As a result, the Siwalik Hills just north of the MFT represent a locus of uplift and erosion, with rates as high as 10 mm/yr (Lavé and Avouac 2001). Relief there is not as dramatic as in the High Range due to the locally high precipitation and runoff combined with poorly consolidated rocks that are easy to erode.

Downstream of the MFT, the land subsides as a result of the push exerted by the Himalaya over the underlying Indian Plate. Subsidence creates "accommodation space" where sediment can be deposited by rivers (Figure 2.1). The heaviest, largest grains are deposited first: all rivers exhibit an abrupt gravel–sand transition (GST, Dingle et al. 2016). Most of the gravel is trapped upstream of the GST, and downstream rivers are sandy. The GST is typically 10–45 km downstream of the MFT, and this distance appears to be controlled by the subsidence rate: the greater the subsidence rate, the greater the space created to store gravel. The GST therefore tends to be closer to the MFT in Central Nepal, where subsidence is fastest (e.g. Kosi River), than in the western Himalaya (Dingle et al. 2016).

It is interesting to think about what controls the amount of gravel and sand delivered to the sedimentary basin (Figure 2.6). Sediment grains originate on the hillslopes or in stocks of glacial sediment. As grains are transported downstream, they are reduced in size by abrasion. It appears that most of the Himalayan gravel originates less than ~100 km of the mountain front (Dingle et al. 2017): grains sourced further away will have been turned into sand by the time they reach the mountain front, except for very resistant rock types such as quartzite that can subsist over longer distances. The importance of abrasion is demonstrated by the fact that quartzite can make up in excess of 90% of the gravel at the mountain front, despite representing ~10% of the rocks exposed in a given catchment (Attal and Lavé 2006; Dingle et al. 2017; Quick et al. 2020).

The implications of these findings are important: earthquakes cause widespread landsliding, supplying rivers with large amounts of sediment. An earthquake causing intense shaking and landsliding in the Higher Himalaya, far away from the mountain front, will likely lead to a pulse of sand to the mountain front, as the bulk of the sediment will have been turned into sand by abrasion by the time it reaches the MFT. Conversely, an earthquake causing intense shaking and landsliding near the mountain front will lead to a pulse of gravel out of the mountains (Dingle et al. 2017). The earthquake location therefore controls the response of the rivers immediately downstream of the mountain front, with potentially very different impacts on the communities that rely on them for their subsistence.

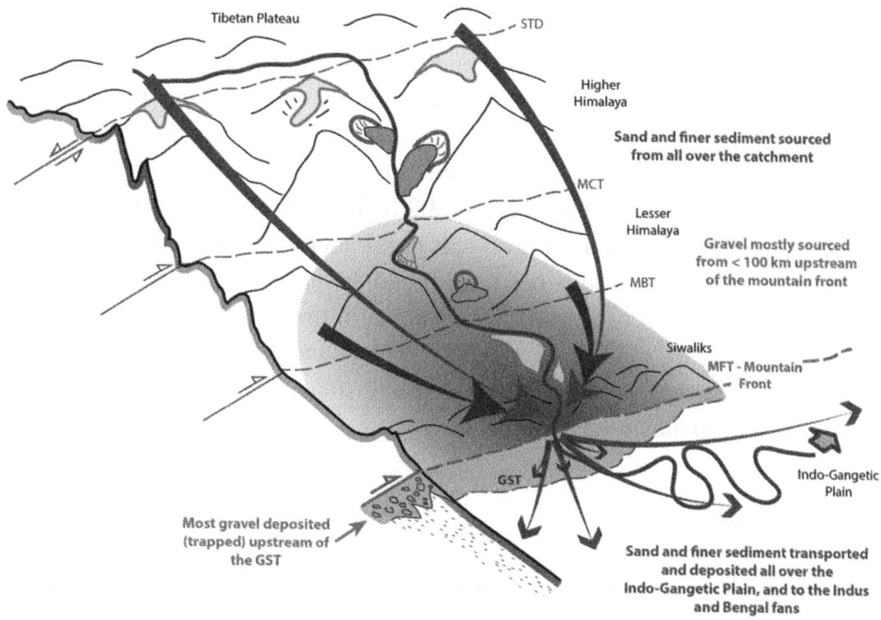

Figure 2.6. *Schematic morpho-tectonic summary of an ideal trans-Himalayan river (see Figure 2.1 for caption) illustrating the dichotomy between the pathways of sand and gravel. Gravel's transport distance is restricted by abrasion and the inability for rivers to transport gravel beyond the gravel–sand transition (GST): most of the gravel exiting the mountains is sourced from up to ∼100 km of the mountain front (except for the hardest rock types such as quartzites that can be transported as gravel for hundreds of kilometers); most gravel is deposited upstream of the GST (up to ∼40 km downstream of the mountain front). Meanwhile, sand and finer sediment originate from everywhere in the catchment, and may be transported downstream for hundreds to thousands of kilometers, all the way to the Indus and Bengal fans. For a color version of this figure, see www.iste.co.uk/cattin/himalaya3.zip*

The presence or absence of gravel in the rivers of the Indo-Gangetic plains also influences their behavior. Upstream of the GST, gravel-bedded rivers are stable over timescales of century to millennia but migrate across their alluvial fans through sudden avulsion, such as the 2008 Kosi avulsion that displaced ∼3 million people (Sinha et al. 2014). Downstream of the GST, rivers actively transport sand and finer sediment across the plains and to the Indus and Bengal fans. The meandering rivers migrate through bank erosion at rates that can exceed hundreds of meters per year, presenting different challenges to the communities living on their floodplain (Dingle et al. 2020).

Sand and finer sediment is regularly exported from the mountains during the monsoon, but gravel export tends to be more episodic. While some of the gravel can be put in motion on a yearly basis during the monsoon peak flow, grains on the coarsest gravel bars require powerful floods to be mobilized; such floods may occur as infrequently as once every few hundreds or thousands of years (Dingle et al. 2020). Evidence suggests that extreme floods can occasionally push gravel tens of kilometers downstream of the GST. Evidence for extreme floods has been found as far as hundreds of kilometers offshore, in the form of large debris in cores in the Bengal Fan. As extremes are exacerbated by climate change, our understanding of rivers' response to storms, earthquakes and landslides can help make the communities of the Indo-Gangetic plains more resilient.

2.7. Conclusion

Quantifying the surface processes at work in mountainous landscapes, such as the Himalaya, is crucial to determine the relationship between tectonics, climate, erosion and topography, and to mitigate the risks and hazards associated with erosion processes. New techniques are assisting researchers in constraining erosion rates over a wide range of temporal and spatial scales. They highlight the connection between various domains of the mountain building systems, and in particular the cascades of hazards that might be triggered. They also highlight the strong feedback between climate (driven by orbital parameters) and erosion rates and sediment flux. These connections operate on a range of temporal and spatial scales. For example, co-seismic shaking leads to widespread landsliding, which in turn leads to increased sediment flux in the river system and often triggers flooding. Oscillating variations in monsoonal strength foster changes from supply-limited to transport-limited conditions, causing cycles of widespread sediment aggradation (with thicknesses of up to >100 m) followed by the re-incision of the fluvial network and removal of the sediment over a few thousand years. Ongoing research on surface processes also demonstrates that extreme events play a major role in the shaping of landscapes and the export of sediment from mountains to basins. Measurement of orogen-wide erosion rates across the Himalaya on various timescales has documented that both the southern part of the High Himalaya (with bedrock incision rates of ~2–5 mm/yr) and the uplifted Siwalik Hills (recycling of Miocene foreland basin sediments of ~5–10 mm/yr) are the most important sediment sources for the Indo-Gangetic Plain as well as the marine Indus and Bengal fans.

Mountain environments are at the forefront of climate change, with widespread glacial retreat and a dramatic increase in the frequency of extreme events. Retreating glaciers are exposing new sediment archives in the high mountains, and generate proglacial lakes behind fragile moraine dams, which can lead to deadly glacial outburst floods. The regular occurrence of devastating precipitation events in the Himalaya such as the 2013 Uttarakhand floods (North India) or the 2021 Melamchi disaster (Central Nepal) demonstrates the need to develop our understanding of erosional systems, to improve the way we assess, monitor and mitigate the risks they pose to increasingly vulnerable populations and costly infrastructure both in the mountains and on the plains.

2.8. References

Attal, M. and Lavé, J. (2006). Changes of bedload characteristics along the Marsyandi River (central Nepal): Implications for understanding hillslope sediment supply, sediment load evolution along fluvial networks, and denudation in active orogenic belts. *Geological Society of America Special Paper: Tectonics, Climate, and Landscape Evolution*, 398, 143–171.

van der Beek, P., Litty, C., Baudin, M., Mercier, J., Robert, X., Hardwick, E. (2016). Contrasting tectonically driven exhumation and incision patterns, western versus central Nepal Himalaya. *Geology*, 44(4), G37579.1.

Blöthe, J.H. and Korup, O. (2013). Millennial lag times in the Himalayan sediment routing system. *Earth and Planetary Science Letters*, 382, 38–46.

Bollinger, L., Henry, P., Avouac, J.P. (2006). Mountain building in the Nepal Himalaya: Thermal and kinematic model. *Earth and Planetary Science Letters*, 244(1–2), 58–71.

Bookhagen, B. and Burbank, D.W. (2010). Toward a complete Himalayan hydrological budget: Spatiotemporal distribution of snowmelt and rainfall and their impact on river discharge. *Journal of Geophysical Research*, 115(F3), F03019.

Bookhagen, B., Thiede, R.C., Strecker, M.R. (2005). Late Quaternary intensified monsoon phases control landscape evolution in the northwest Himalaya. *Geology*, 33(2), 149–152.

Bookhagen, B., Fleitmann, D., Nishiizumi, K., Strecker, M., Thiede, R.C. (2006). Holocene monsoonal dynamics and fluvial terrace formation in the northwest Himalaya, India. *Geology*, 34(7), 601–604.

Brozovic, N., Burbank, D.W., Meigs, A.J. (1997). Climatic limits on landscape development in the northwestern Himalaya. *Science*, 276(5312), 571–574.

Burbank, D.W., Leland, J., Fielding, E., Anderson, R.S., Brozovic, N., Reid, M.R., Duncan, C. (1996). Bedrock incision, rock uplift and threshold hillslopes in the northwestern Himalayas. *Nature*, 379(6565), 505–510.

Chirouze, F., Huyghe, P., Chauvel, C., van der Beek, P., Bernet, M., Mugnier, J.-L. (2015). Stable drainage pattern and variable exhumation in the western Himalaya since the Middle Miocene. *The Journal of Geology*, 123(1), 1–20.

Clift, P.D., Hodges, K.V., Heslop, D., Hannigan, R., Van Long, H., Calves, G. (2008). Correlation of Himalayan exhumation rates and Asian monsoon intensity. *Nature Geoscience*, 1(12), 875–880.

Codilean, A.T., Munack, H., Cohen, T.J., Saktura, W.M., Gray, A., Mudd, S.M. (2018). OCTOPUS: An open cosmogenic isotope and luminescence database. *Earth System Science Data*, 10(4), 2123–2139.

Cook, K.L., Andermann, C., Gimbert, F., Adhikari, B.R., Hovius, N. (2018). Glacial lake outburst floods as drivers of fluvial erosion in the Himalaya. *Science*, 362(6410), 53–57.

Dahlquist, M.P. and West, A.J. (2022). The imprint of erosion by glacial lake outburst floods in the topography of central Himalayan rivers. *Earth Surf. Dynam.*, 10, 705–722. https://doi.org/10.5194/esurf-10-705-2022 [Online]. Available at: https://esurf.copernicus.org/articles/10/705/2022/.

Deal, E., Favre, A.-C., Braun, J. (2017). Rainfall variability in the Himalayan orogen and its relevance to erosion processes. *Water Resources Research*, 53(5), 4004–4021.

Dettman, D.L., Kohn, M.J., Quade, J., Ryerson, F., Ojha, T.P., Hamidullah, S. (2001). Seasonal stable isotope evidence for a strong Asian monsoon throughout the past 10.7 m.y. *Geology*, 29(1), 31–34.

Dey, S., Bookhagen, B., Thiede, R.C., Wittmann, H., Chauhan, N., Jain, V., Strecker, M.R. (2022). Impact of Late Pleistocene climate variability on paleo-erosion rates in the western Himalaya. *Earth and Planetary Science Letters*. https://doi.org/10.1016/j.epsl.2021.117326.

Ding, L., Spicer, R., Yang, J., Xu, Q., Cai, F., Li, S., Lai, Q., Wang, H., Spicer, T., Yue, Y. et al. (2017). Quantifying the rise of the Himalaya orogen and implications for the South Asian monsoon. *Geology*, 45(3), 215–218.

Dingle, E.H., Sinclair, H.D., Attal, M., Milodowski, D.T., Singh, V. (2016). Subsidence control on river morphology and grain size in the Ganga Plain. *American Journal of Science*, 316(8), 778–812.

Dingle, E.H., Attal, M., Sinclair, H.D. (2017). Abrasion-set limits on Himalayan gravel flux. *Nature*, 544(7651), 471–474.

Dingle, E.H., Sinclair, H.D., Venditti, J.G., Attal, M., Kinnaird, T.C., Creed, M., Quick, L., Nittrouer, J.A., Gautam, D. (2020). Sediment dynamics across gravel-sand transitions: Implications for river stability and floodplain recycling. *Geology*, 48(5), 468–472.

Fick, S.E. and Hijmans, R.J. (2017). WorldClim 2: New 1-km spatial resolution climate surfaces for global land areas. *International Journal of Climatology*, 37(12), 4302–4315.

Gabet, E.J., Burbank, D.W., Putkonen, J.K., Pratt-Sitaula, B.A., Ojha, T. (2004a). Rainfall thresholds for landsliding in the Himalayas of Nepal. *Geomorphology*, 63(3–4), 131–143.

Gabet, E.J., Pratt-Sitaula, B.A., Burbank, D.W. (2004b). Climatic controls on hillslope angle and relief in the Himalayas. *Geology*, 32(7), 629–632.

Ghoshal, S., McQuarrie, N., Robinson, D.M., Adhikari, D.P., Morgan, L.E., Ehlers, T.A. (2020). Constraining Central Himalayan (Nepal) fault geometry through integrated thermochronology and thermokinematic modeling. *Tectonics*, 39(9), 1–33.

Godard, V. and Tucker, G.E. (2021). Influence of climate-forcing frequency on hillslope response. *Geophysical Research Letters*, 48(18), 1–11.

Godard, V., Bourles, D.L., Spinabella, F., Burbank, D.W., Bookhagen, B., Fisher, G.B., Moulin, A., Leanni, L. (2014). Dominance of tectonics over climate in Himalayan denudation. *Geology*, 42(3), 243–246.

Goren, L. (2016). A theoretical model for fluvial channel response time during time-dependent climatic and tectonic forcing and its inverse applications. *Geophysical Research Letters*, 43(20), 10,753–10,763.

Herman, F., Copeland, P., Avouac, J.-P., Bollinger, L., Mahéo, G., Le Fort, P., Rai, S., Foster, D., Pêcher, A., Stüwe, K. et al. (2010). Exhumation, crustal deformation, and thermal structure of the Nepal Himalaya derived from the inversion of thermochronological and thermobarometric data and modeling of the topography. *Journal of Geophysical Research*, 115(B6), B06407.

Huber, M.L., Lupker, M., Gallen, S.F., Christl, M., Gajurel, A.P. (2020). Timing of exotic, far-traveled boulder emplacement and paleo-outburst flooding in the central Himalayas. *Earth Surface Dynamics*, 8(3), 769–787.

Konstantinovskaia, E. and Malavieille, J. (2005). Erosion and exhumation in accretionary orogens: Experimental and geological approaches. *Geochemistry, Geophysics, Geosystems*, 6(2), Q02006.

Lang, K.A., Huntington, K.W., Montgomery, D.R. (2013). Erosion of the Tsangpo Gorge by megafloods, eastern Himalaya. *Geology*, 41(9), 1003–1006.

Larsen, I.J. and Montgomery, D.R. (2012). Landslide erosion coupled to tectonics and river incision. *Nature Geoscience*, 5(7), 468–473.

Lavé, J. and Avouac, J.P. (2000). Active folding of fluvial terraces across the Siwaliks Hills, Himalayas of central Nepal. *Journal of Geophysical Research*, 105(B3), 5735–5770.

Lavé, J. and Avouac, J.P. (2001). Fluvial incision and tectonic uplift across the Himalayas of central Nepal. *Journal of Geophysical Research*, 106(B11), 26561–26591.

Le Roux-Mallouf, R., Godard, V., Cattin, R., Ferry, M., Gyeltshen, J., Ritz, J.-F., Drupka, D., Guillou, V., Arnold, M., Aumaître, G. et al. (2015). Evidence for a wide and gently dipping Main Himalayan Thrust in western Bhutan. *Geophysical Research Letters*, 42(9), 3257–3265.

Lenard, S.J., Lavé, J., France-Lanord, C., Aumaître, G., Bourlès, D.L., Keddadouche, K. (2020). Steady erosion rates in the Himalayas through late Cenozoic climatic changes. *Nature Geoscience*, 13(6), 448–452.

Li, G., West, A.J., Densmore, A.L., Jin, Z., Parker, R.N., Hilton, R.G. (2014). Seismic mountain building: Landslides associated with the 2008 Wenchuan earthquake in the context of a generalized model for earthquake volume balance. *Geochemistry, Geophysics, Geosystems*, 15, 833–844.

Lisiecki, L.E. and Raymo, M.E. (2005). A Pliocene–Pleistocene stack of 57 globally distributed benthic d18O records. *Paleoceanography*, 20(1), PA1003.

Mandal, S.K., Scherler, D., Wittmann, H. (2021). Tectonic accretion controls erosional cyclicity in the Himalaya. *AGU Advances*, 2(3), 1–27.

Marc, O., Behling, R., Andermann, C., Turowski, J.M., Illien, L., Roessner, S., Hovius, N. (2019). Long-term erosion of the Nepal Himalayas by bedrock landsliding: The role of monsoons, earthquakes and giant landslides. *Earth Surface Dynamics*, 7(1), 107–128.

Märki, L., Lupker, M., France-Lanord, C., Lavé, J., Gallen, S., Gajurel, A.P., Haghipour, N., Leuenberger-West, F., Eglinton, T. (2021). An unshakable carbon budget for the Himalaya. *Nature Geoscience*, 14, 745–750. https://doi.org/10.1038/s41561-021-00815-z.

Naylor, M. and Sinclair, H. (2007). Punctuated thrust deformation in the context of doubly vergent thrust wedges: Implications for the localization of uplift and exhumation. *Geology*, 35(6), 559.

Owen, L.A. and Dortch, J.M. (2014). Nature and timing of Quaternary glaciation in the Himalayan-Tibetan orogen. *Quaternary Science Reviews*, 88, 14–54.

Perron, J.T. (2017). Climate and the pace of erosional landscape evolution. *Annual Review of Earth and Planetary Sciences*, 45(1). DOI: 10.1146/annurev-earth-060614-105405.

Quick, L., Sinclair, H., Attal, M., Singh, V. (2020). Conglomerate recycling in the Himalayan foreland basin: Implications for grain size and provenance. *GSA Bulletin*, 132(7–8), 1639–1656.

Roback, K., Clark, M.K., West, A.J., Zekkos, D., Li, G., Gallen, S.F., Chamlagain, D., Godt, J.W. (2017). The size, distribution, and mobility of landslides caused by the 2015 Mw7.8 Gorkha earthquake, Nepal. *Geomorphology*, 301, 121–138.

Robert, X., van der Beek, P., Braun, J., Perry, C., Dubille, M., Mugnier, J.-L. (2009). Assessing Quaternary reactivation of the Main Central thrust zone (central Nepal Himalaya): New thermochronologic data and numerical modeling. *Geology*, 37(8), 731–734.

Robert, X., van der Beek, P., Braun, J., Perry, C., Mugnier, J.-L. (2011). Control of detachment geometry on lateral variations in exhumation rates in the Himalaya: Insights from low-temperature thermochronology and numerical modeling. *Journal of Geophysical Research*, 116(B5), B05202.

Scherler, D., Bookhagen, B., Strecker, M.R. (2014). Tectonic control on 10 Be derived erosion rates in the Garhwal Himalaya, India. *Journal of Geophysical Research: Earth Surface*, 119(2), 1–23.

Scherler, D., DiBiase, R.A., Fisher, G.B., Avouac, J.-P. (2017). Testing monsoonal controls on bedrock river incision in the Himalaya and Eastern Tibet with a stochastic-threshold stream power model. *Journal of Geophysical Research: Earth Surface*, 122(7), 1389–1429.

Schwanghart, W., Bernhardt, A., Stolle, A., Hoelzmann, P., Adhikari, B.R., Andermann, C., Tofelde, S., Merchel, S., Rugel, G., Fort, M. (2015). Repeated catastrophic valley infill following medieval earthquakes in the Nepal Himalaya. *Science*, 351(December), 147–150.

Seeber, L. and Gornitz, V. (1983). River profiles along the Himalayan arc as indicators of active tectonics. *Tectonophysics*, 92(4), 335–367.

Sinha, R., Sripriyanka, K., Jain, V., Mukul, M. (2014). Avulsion threshold and planform dynamics of the Kosi River in north Bihar (India) and Nepal: A GIS framework. *Geomorphology*, 216, 157–170.

Struck, M., Andermann, C., Hovius, N., Korup, O., Turowski, J.M., Bista, R., Pandit, H.P., Dahal, R.K. (2015). Monsoonal hillslope processes determine grain size-specific suspended sediment fluxes in a trans-Himalayan river: Mass wasting determines sediment caliber. *Geophysical Research Letters*, 42(7), 2302–2308.

Thiede, R.C. and Ehlers, T.A. (2013). Large spatial and temporal variations in Himalayan denudation. *Earth and Planetary Science Letters*, 371–372, 278–293.

Thiede, R.C., Robert, X., Stübner, K., Dey, S., Faruhn, J. (2017). Sustained out-of-sequence shortening along a tectonically active segment of the Main Boundary thrust: The Dhauladhar Range in the northwestern Himalaya. *Lithosphere*, 9(5), 715–725.

Vannay, J.-C., Grasemann, B., Rahn, M., Frank, W., Carter, A., Baudraz, V., Cosca, M. (2004). Miocene to Holocene exhumation of metamorphic crustal wedges in the NW Himalaya: Evidence for tectonic extrusion coupled to fluvial erosion: Tectonics and erosion in the NW Himalaya. *Tectonics*, 23(1).

Veh, G., Korup, O., von Specht, S., Roessner, S., Walz, A. (2019). Unchanged frequency of moraine-dammed glacial lake outburst floods in the Himalaya. *Nature Climate Change*, 9(5), 379–383.

Vögeli, N., Huyghe, P., van der Beek, P., Najman, Y., Garzanti, E., Chauvel, C. (2018). Weathering regime in the Eastern Himalaya since the mid-Miocene: Indications from detrital geochemistry and clay mineralogy of the Kameng River Section, Arunachal Pradesh, India. *Basin Research*, 30(1), 59–74.

West, A.J., Arnold, M., Aumaître, G., Bourlès, D.L., Keddadouche, K., Bickle, M., Ojha, T. (2015). High natural erosion rates are the backdrop for present-day soil erosion in the agricultural Middle Hills of Nepal. *Earth Surface Dynamics*, 3(3), 363–387.

Whipple, K.X. and Meade, B.J. (2006). Orogen response to changes in climatic and tectonic forcing. *Earth and Planetary Science Letters*, 243(1–2), 218–228.

Zachos, J.C., Dickens, G.R., Zeebe, R.E. (2008). An early Cenozoic perspective on greenhouse warming and carbon-cycle dynamics. *Nature*, 451(7176), 279–283.

PART 2

Natural Hazards

3

Glaciers and Glacier Lake Outburst Floods in the Himalaya

Christoff ANDERMANN[1], Santosh NEPAL[2], Patrick WAGNON[3], Georg VEH[4], Sudan Bikash MAHARJAN[2], Mohd Farooq AZAM[5], Fanny BRUN[3] and Wolfgang SCHWANGHART[4]

[1] *GFZ, Potsdam, Germany*

[2] *International Centre for Integrated Mountain Development, Kathmandu, Nepal*

[3] *University of Grenoble Alpes, France*

[4] *University of Potsdam, Germany*

[5] *Indian Institute of Technology Indore, Simrol, India*

3.1. Introduction

The glaciers in the Himalaya award the highest mountain range with their majestic visage. The mountain range, which includes the Himalaya and the Karakoram, is more than 2,500 km long (E–W), and hosts the largest area of permanent ice cover outside the polar regions (Figure 3.1). The snow-white peaks are not only an attraction for mountain tourists but also

Himalaya, Dynamics of a Giant 3,
coordinated by Rodolphe CATTIN and Jean-Luc EPARD.

provide valuable water resources for the downstream regions, coining the Himalaya the Asian water towers (Immerzeel et al. 2020; Viviroli et al. 2020). Together with groundwater (Andermann et al. 2012), glaciers are an important water storage that maintains the water supply through the long-lasting dry season (Biemans et al. 2019; Azam et al. 2021). In particular at the syntaxes, glaciers and snow are the major water reservoirs and important for domestic water use, agriculture and hydropower production. Throughout the Himalaya and their downstream nations, people rely on mountain-sourced waters and their associated ecosystem services (e.g. Bookhagen and Burbank 2010; Immerzeel and Bierkens 2012; Pritchard 2019; Viviroli et al. 2020). The Intergovernmental Panel on Climate Change (IPCC) sees these resources at risk, particularly in the central and eastern parts of the Himalaya (Hock et al. 2019). In contrast, the western Himalayan region and Karakoram seem to have experienced increased moisture supply and snowfall, counterbalancing the warming and associated melting (Hewitt 2005; Azam et al. 2018; Farinotti et al. 2020; Kok et al. 2020).

Besides their importance for water resources, glaciers in the Himalaya are associated with natural hazards including glacier lake outburst floods (GLOFs), ice brake off, avalanches, slope failures due to glacial erosion and glacial retreat, and destabilization of partially frozen debris (Ghimire 2004; Veh et al. 2020; Shugar et al. 2021). Climate warming is likely to increase the frequency and magnitude of these processes in the future (e.g. Hugonnet et al. 2021) because glacial retreat will create unstable landforms (see Volume 3 – Chapter 4) and promote enhanced sediment entrainment. Once mobilized and routed downstream, these sediments can create new problems. River aggradation may bury arable lands and increase the risk of inundation. High sediment concentrations also impact on hydropower production due to reservoir siltation and turbine abrasion (Schwanghart et al. 2016, 2018). One of the most immediate effects of glacier retreat is the development of glacial lakes and increased likelihood of GLOFs which have the potential to harm downstream communities, infrastructure and livelihood-related activities. The glacial lakes are located very high in the landscape (often >4,000 masl) and can release huge amounts of water ($> 10^5$ m^3) in a short time. The lakes are characterized by an unratable dam of ice and glacial debris at the glacier front or within the glacier itself.

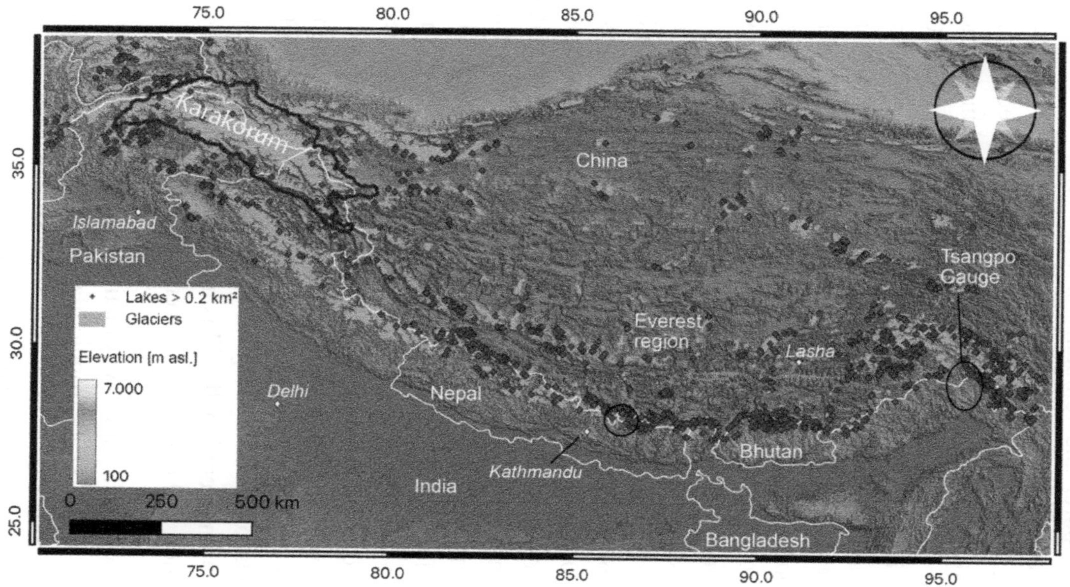

Figure 3.1. *Glacier and glacial lake map of the Himalayan region. Lakes subsampled for lake area provided by the National Cryosphere Desert Data Center (http://www.ncdc.ac.cn) (Wang et al. 2020b) subsampled to lake area > 2 km^2 (1,793 lakes). Glaciers mapped from the Randolph Glacier Inventory 6.0 (https://www.glims.org/RGI/rgi60_dl.html and Pfeffer et al. 2014). The Karakoram region is outlined in black. A particular point of interests is indicated with arrows and names. Elevation is an SRTM-DEM 90 m (http://srtm.csi.cgiar.org). For a color version of this figure, see www.iste.co.uk/cattin/himalaya3.zip*

GLOFs are notoriously difficult to assess and predict. Their unpredicted occurrence, sheer destructive power and relatively short-lived perseverance make it very difficult to obtain reliable observations and measurements of these events (Cook et al. 2018). In almost all cases, scientists are bound to rely on witness reports and remnants of the event, such as transported boulders or high-water marks (e.g. Cenderelli and Wohl 2001; Lang et al. 2013; Huber et al. 2020). All sensors and instruments located in the river and proximity are typically destroyed during the event, resulting in no existing record of the flood wave and flood propagation for historical GLOFs. Only recent innovations in the field of environmental seismology have provided novel insights into so far two events (Cook et al. 2018, 2021). GLOFs destroy everything on their way downstream and what is within the reach of the flood wave. Hydropower facilities along the rivers, for example, are always at risk, as well as roads, settlements and farmland. The erosive power of the floods also leads to undercutting of adjacent slopes and consequently activation of hillslope processes such as landsliding. These hillslope activations take much longer than the channel-related processes themselves and might cause delayed effects by propagating uphill over years to decades. The recent years of research on GLOFs and their consequences have shown the importance of GLOF events in the formation of high mountain landscapes (Dahlquist and West 2021). They not only transport large amounts of materials through the landscape but also have the power to mobilize the river bed and all the clasts located there (e.g. Huber et al. 2020). This is typical because in the headwaters GLOFs exceed rain-born floods by several orders of magnitudes (see Volume 3 – Chapter 2) and thus can mobilize boulders that exceed the transport capacities of other events. Early warning of GLOFs is due to the high elevation and difficult terrain challenging. Often, the sensing technology suffers under fierce weather conditions and warning infrastructure, such as sirens, is costly and tedious to maintain operationally. This unrobust early warning technology adds to the destruction and makes these events even deadlier.

In this chapter, we will guide systematically from the ice-covered peaks and the glacial lakes along a catastrophic glacier lake outburst flood trajectory, through the river channel network to the downstream living societies. The reader will learn about glacier dynamics and the link between GLOF hazards and societal impacts. We will discuss the current state of knowledge with a special focus on climatic changes and their potential consequences. We will demonstrate this with well-documented examples from different places around the Himalaya. The chapter will be wrapped up with an outlook on future potential early warning systems and hazard mitigation with respect to climatic change.

3.2. Glaciers and their future

3.2.1. *The present state of glaciers at a regional scale*

Estimates of the area covered by Himalayan glaciers range between 37 and 51×10^3 km^2 (Figure 3.1), yielding an estimated ice volume between 2,935 and 4,737 km^3 (Pfeffer et al. 2014; Azam et al. 2018). Herreid and Pellicciotti (2020) estimated that 13.7% and 18.7% of the glacierized area is covered by debris, depending on whether they are located in the western or eastern part of the range, respectively. This surface has been increasing in the eastern part in the recent years in relation to shrinking glaciers (Scherler et al. 2018). In a thorough synopsis, Azam et al. (2018) collected and compiled all published observations, both in situ and from satellite, in this region. Their compilation contains 154 historical records of glacier length or area variations, some of which date back to 1840. Still, only 24 glaciers are (or were) monitored for mass balance measurements. These records are generally discontinuous, sometimes of poor quality and rather short since the first measurement began in 1974. Fortunately, developments in Earth observation techniques have helped to close this gap and made it possible to quantify mass variations of glaciers in the entire region since the 1960s (Brun et al. 2017; Maurer et al. 2019).

Glacier evolution is not homogeneous across the region (Figure 3.2). In the Himalaya, there has been a widespread retreat of the glacier fronts, since the mid-19th century, with high variability across glaciers and sub-regions. In the Karakoram, without considering those that are surging, glaciers have remained stable or have advanced slightly (Azam et al. 2018). Glacier mass balance is a more relevant variable, because it is directly related to climate. The mass balance of all glaciers in the Himalaya has only been known comprehensively since the 2000s, and incompletely for the period 1960s–2000s, often by the geodetic method (difference in digital terrain models built from satellite images). The most recent study (Hugonnet et al. 2021) shows that glaciers in the high mountains of Asia lost mass at a rate of -0.22 ±0.05 m w.e./yr (w.e.: water equivalent) between 2000 and 2019, about two times less than the global average over the same period for glaciers. However, there are large disparities between regional mass balances, which range between -0.14 ±0.05 m w.e./yr in the west and -0.47 ±0.05 m w.e./yr in the east of the Himalaya. Furthermore, for these regions, mass loss appears to have accelerated between the beginning and end of the last two decades with approximately a doubling of loss in 2015–2019 compared to 2000–2004.

Brun et al. (2017) and Shean et al. (2020) quantified region-wide mass balances across the high mountains of Asia between 2000 and 2016, and between 2000 and 2018, respectively, with values for 12 sub-regions (Figure 3.2). Glacial mass losses are large across the whole Himalaya and maximum in its south-eastern part (-0.55 ±0.13 m w.e./yr between 2000 and 2018). It is also interesting to note that glaciers that terminate in lakes lose more mass than the land terminating glaciers (Brun et al. 2019; Maurer et al. 2019).

In the Karakoram (Figure 3.1), on the contrary, glaciers are almost stable (-0.04 ±0.04 m w.e./yr between 2000 and 2018). This stability of the Karakoram glaciers, unique on our planet, has been known since 2005 and termed "Karakoram anomaly" (Hewitt 2005). This anomaly is in fact centered on the region immediately north-east of the Karakoram, the Western Kun Lun, where glaciers are slightly gaining mass (+0.04±0.04 m w.e./yr between 2000 and 2018). Thanks to the declassified spy satellite imagery from the 1960s, the same kind of study could be done, but less accurately and only over limited areas. The regional disparities observed since 2000 existed since the 1960s, although there was an intensification of mass loss after 2000 compared to previous decades (Maurer et al. 2019). The Karakoram anomaly has existed at least since the 1970s (Bolch et al. 2017) but it seems to be ending (Hugonnet et al. 2021).

Farinotti et al. (2020) explored possible reasons for the Karakoram anomaly. They showed that Karakoram's glaciers have a unique morphology linked to the extreme topography of the region, a dense debris cover and a high avalanche activity. Moreover, there is an increase in annual winter and summer precipitation combined with a decrease in average and minimum temperatures during summer (Archer and Fowler 2004; Fowler and Archer 2006). Several hypotheses have been put forward to explain these climate trends, from a change in the large-scale westerly circulations bringing precipitation (Forsythe et al. 2017), which have recently shifted towards the Karakoram, to an intensification of the water cycle on a regional scale, due to the development of irrigation in the Tarim Basin located on the north-east of the Karakoram (Kok et al. 2018).

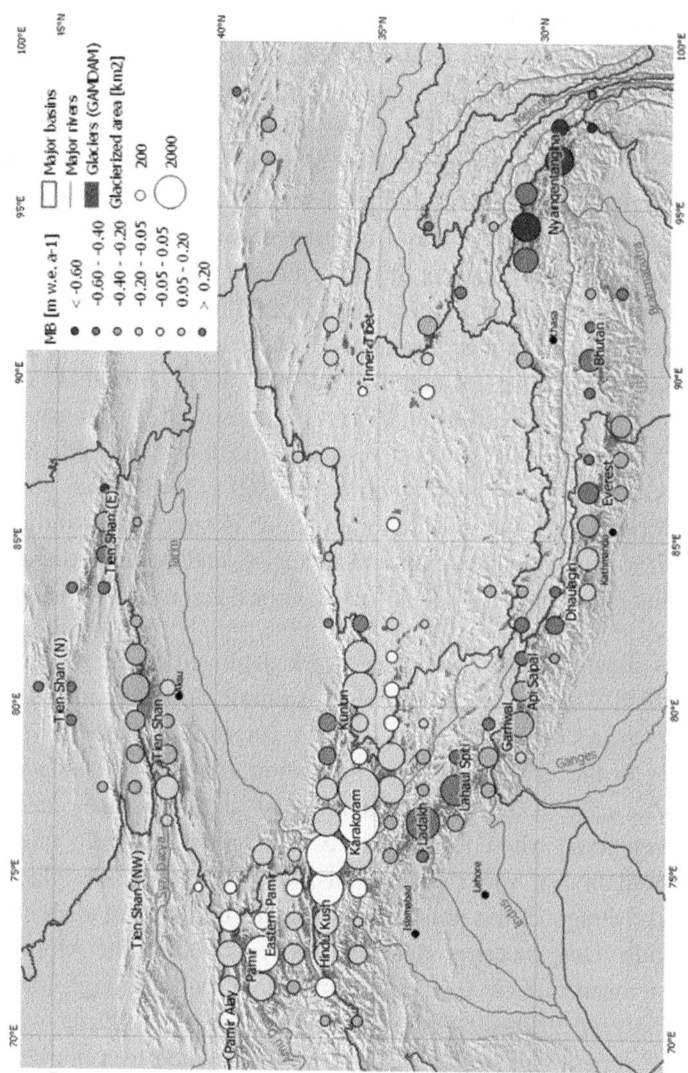

Figure 3.2. *Map of glacier mean elevation change on a 1° grid for the period 2000–2016. The dots are color-coded by the rate of glacier elevation change. The area of the dots is proportional to the glacierized area. Figure from (Brun et al. 2017). For a color version of this figure, see www.iste.co.uk/cattin/himalaya3.zip*

3.2.2. *On the interest of monitoring glaciers*

Although satellite images have led to an immense progress in quantifying the evolution of glacier volumes over the entire Himalaya–Karakoram since 2000, these data are still insufficient to provide insight into the interannual or seasonal variability of mass balance. Thus, to understand the controls on glacier extents, in situ monitoring of glaciers is and will remain indispensable. Field-surveyed glaciers are carefully chosen to be representative of various climatic zones. Together with glaciological approaches to glacier mass balance (e.g. with the help of snow stakes and pits), researchers continuously monitor the glacial environment by measuring meteorological variables such as temperature, precipitation and humidity to determine the processes that control the accumulation or ablation of snow and ice, and to estimate the processes' sensitivity to climatic variables. Examples of such detailed monitoring efforts are the Chhota Shigri glacier located in Northern India in Himachal Pradesh or the Mera glacier, not far from Mount Everest, whose series of mass balances initiated in 2002 and 2007, respectively, are the longest in the Himalaya and Nepal (Azam et al. 2016; Wagnon et al. 2021). To date, no continuous glacier monitoring exists in the Karakoram which is largely due to the lack of long-term investments, both human and financial, and because these remote and high-altitude regions are difficult to access or dangerous to work in.

Figure 3.3 shows the observation network deployed on the Mera Glacier in Nepal, which extends from the Mera summit at 6,500 m a.s.l. to its front at 4,900 m a.s.l.. The mass balance is determined twice a year from measurements in the ablation and accumulation zone. In the ablation zone below 5,600 m a.s.l., a dense, regular array of ablation stakes are used to measure emergences, whereas snow core drilling devices help quantifying rates of accumulation at various elevations, up to the top of the glacier. The area-weighted average of the point mass balances obtained at each site gives the glacier-wide mass balance. Between 2007 and 2020, the Mera Glacier lost mass, at a mean rate of -0.42 ±0.20 m w.e./yr, but with high interannual variability. The year 2017–2018 showed the most negative glacier-wide mass balance of the series with -0.92 ±0.16 m w.e./yr, related to a marked deficit in precipitation, while three years were characterized by a slight mass gain, largely related to record-breaking amounts of precipitation as those emplaced during the passage of Typhoon Phailin between October 13 and 15, 2013 (Wagnon et al. 2021). In addition to the glaciological measurements, several meteorological stations are continuously operating next to or on the surface of the glacier itself. These have shown that glacier melt is overwhelmingly controlled by incoming solar and infrared atmospheric radiation, hence the importance of clouds, and snowfall influencing the surface albedo.

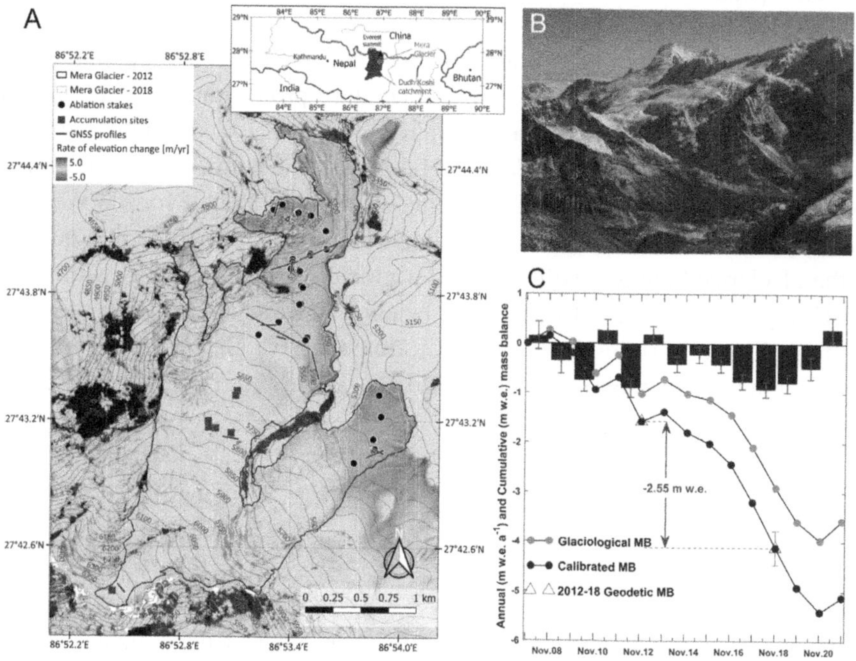

Figure 3.3. *Monitoring Himalayan glaciers. (A) Example of glacier monitoring (Mera Glacier – Khumbu area – Nepal) showing a map with the monitoring network on the glacier. (B) Photo of Mera Glacier where the monitoring is installed. (C) Elevation changes between 2012 and 2018, and the annual and cumulative glacier-wide mass balance between 2007 and 2021. Adapted and updated from Wagnon et al. (2021). For a color version of this figure, see www.iste.co.uk/cattin/himalaya3.zip*

3.2.3. *What will happen to the Himalayan glaciers and Karakoram glaciers?*

With the recent climate change impact on the cryosphere, the future of glaciers is a matter of concern, especially in the high mountains of Asia where about 800 million people partly depend on water that comes from glaciers for drinking water, irrigation or hydropower (Pritchard 2019). Kraaijenbrink et al. (2017) conducted the most comprehensive study to estimate future losses from glaciers in the high mountains of Asia, taking into account elevation-dependent warming, and processes specific to these glaciers such as their debris cover, or the presence of supraglacial lakes. They conclude that by 2100, depending on the climate scenario, between one-third and two-thirds of current ice volumes will disappear, with strong disparities between regions. Kraaijenbrink et al. (2017) projected the evolution of glaciers in the high mountains of Asia, assuming global warming is contained to 1.5°C by 2100, as recommended by the Paris agreement of 2015. The temperature projection shows amplified warming with altitude, and reaches values of 1.9–2.3°C depending on the sub-region. In the Central and Eastern Himalaya, two-thirds of the glacierized area are expected to be lost by 2100. This would only be half for the Western Himalaya. The Karakoram is the least impacted region, but a loss of a quarter of the glacierized areas is still expected if global warming remains within the 1.5°C target. However, with the world heading towards an emission scenario pushing climate warming beyond 1.5°C, more massive glacier losses must be expected in the Himalaya and in the world as a whole (e.g. Potocki et al. 2022).

3.3. Glacier lakes

3.3.1. *Formation and present distribution of glacier lakes in the Himalaya*

Phases of advance and retreat of Himalayan glaciers have created space for lakes to grow at the glacial margins or forelands (Figure 3.4). Glacier lakes retain snow and glacier meltwater behind glacial ice, moraines or bedrock. Other lakes exist beneath (subglacial), within (englacial) or on top (supraglacial) of glaciers. Glacier lakes are important buffers in the hydrological cascade, trapping sediments, nutrients and meltwater (Carrivick and Tweed 2013). With the ongoing glacier recession, glacier lakes could gradually take over the role of glaciers as important reservoirs of freshwater storage in high mountains. People living downstream of mountains could thus benefit from glacier lakes as a reliable source for drinking water, irrigation and hydropower production (Farinotti et al. 2019; Pritchard 2019; Immerzeel et al. 2020).

These potential benefits contrast with the hazard that glacier lakes pose to people and infrastructure, when they drain suddenly without warning. In the

Himalaya, GLOFs have repeatedly inflicted losses, with Bhutan and Nepal suffering the greatest economic damage from GLOFs worldwide (Carrivick and Tweed 2016).

Figure 3.4. *Photographs of glacier lakes. A) End moraine deposits and lower Barun glacial lake (photo: Sudan Bikash Maharjan, 2021). B) Cascading glacial lake formation on the glacial moraine deposits, Makalu Barun National Park, Nepal (photo: Sudan Bikash Maharjan, 2021). C) Tsho Rolpa lake, the largest glacial lake in Nepal (photo: Tika Ram Gurung). D and E) Gangapurna lake in Manang (photo: Nabin Baral). F) Barun glacial lake (Moraine dammed) at Makalu base camp, Nepal (photo: Sudan Bikash Maharjan, 2021). For a color version of this figure, see www.iste.co.uk/cattin/himalaya3.zip*

Mapping glacier lakes, assessing their geometric and topographic properties, and tracking their changes aim to better quantify GLOF hazard, thus supporting sustainable development in mountain valleys (Mool et al. 2001; Gardelle et al. 2011; Zheng et al. 2021). The freely available satellite imagery, particularly from the Landsat mission, has been very helpful to detect lakes throughout the mountain range, while high-resolution Google Earth imagery allows us to determine the type of dam that impounds these lakes. Accordingly, a census on manually mapped lake extents in 2005 suggests that the diversity, abundance and area of lakes vary regionally, reflecting differences in the maximum (pre)historical glacier extent and deglaciation since then (Maharjan et al. 2018).

Today, the Himalaya host >25,000 glacier lakes (with lake areas $\geq$ 3,000 m^2) with a total area of ~1,450 km^2 (Maharjan et al. 2018). In terms of abundance, most of the lakes occur in the upper reaches of the Brahmaputra River, which drains the eastern Himalaya and the Hengduan Shan. Almost two-thirds of all lakes are impounded by bedrock, and this type of impoundment also prevails in most river basins that originate in the Himalaya (Maharjan et al. 2018). Such bedrock-dammed lakes are located in glacial cirques or valleys that were eroded during earlier glacier advances and left behind as water-filled overdeepenings. Moraine-dammed lakes (see also Figure 3.4) account for one-third of the Himalayan lake abundance, with regional clusters in the central and eastern Himalaya (Ganges and Brahmaputra basins). A small proportion of the Himalayan lakes (5%) are impounded by glacier dams, locally confined to a few valleys in the Hindu Kush and Karakoram (Iturrizaga 2005b, 2005b). Glacier-dammed lakes can form when a glacier advances (surges) from a tributary valley into the main valley and temporarily blocks river discharge.

Episodes of valley blockages can occur cyclically (Bazai et al. 2021). For example, the Kyagar, Khurdopin and Shishper glaciers in the upper Indus basin form dams about every 15–20 years (Bazai et al. 2021). In contrast to lateral damming, supraglacial lakes form on the glacier surface that is often covered with a layer of debris (Benn et al. 2012). The number of supraglacial lakes is difficult to determine precisely at any given time because these lakes can fill with meltwater and empty seasonally (Miles et al. 2017). The spatial resolution of satellite sensors further plays an important role in identifying these waterbodies in optical imagery. The 2005 lake census based on Landsat imagery at a 30-meter resolution found > 1,200 supraglacial lakes throughout the Himalaya (Maharjan et al. 2018). Recent studies found a higher count on only seven selected glaciers in the Karakoram and Everest regions because they used Planet or Pleiades images with higher spatial resolution (< 3 m) (Watson et al. 2018; Wendleder et al. 2021). Subglacial and englacial lakes form in cavities and veins of the glacier. These lakes largely elude detection with optical satellite imagery, so their existence only occasionally becomes known when they empty (Livingstone et al. 2022).

About 90% of all lakes have areas < 0.1 km^2. Yet, the largest lakes contribute disproportionately to the total lake area: only 122 or 0.5% of all lakes (0.5% of all) have areas >1 km^2, but these lakes account for 20% of the total lake area (Maharjan et al. 2018). Lakes cover more area in the eastern and central parts of the Himalaya, drained by the Ganges and Brahmaputra, than in the northern regions. Some of the largest lakes are located in the border region between Bhutan, Nepal and China, are dammed by end moraines or bedrock, and can have volumes of > 108 m^3 (Wang et al. 2015; Haritashya et al. 2018; Rinzin et al. 2021). Though smaller in abundance, glacier-dammed lakes in the Karakoram can attain similar volumes (Round et al. 2017; Bazai et al. 2021). Supraglacial lakes have the smallest average size among all lakes and rarely exceed an area of 0.01 km^2 (Maharjan et al. 2018). However, subaqueous and subaerial melting of adjacent ice cliffs and increased calving rates can cause supraglacial lakes to expand and coalesce into larger moraine-dammed lakes (Benn et al. 2012). For example, between 2001 and 2010, lakes on the Ngozumpa glacier grew via spillways to a lake of 0.3 km^2 with average annual growth rates of 10% (Thompson et al. 2012).

3.3.2. *Historic changes of glacier lake abundance and size*

The distribution of glacier lakes in the Himalaya has changed rapidly due to enhanced meltwater input from retreating glaciers in past decades. Scientists have been particularly concerned with settings where glaciers terminate in overdeepened depressions and moraines. When these glaciers retreat from their advanced positions, lakes can rapidly form and grow as these basins become ice-free. Once a lake has formed, warm water temperatures and the exposure of the lake to wind can promote thermal undercutting of the glacier front, and these processes increase in efficiency as the water surface becomes larger (Fujita et al. 2009). Lake-terminating glaciers in the Central and Eastern Himalaya were found to have higher mass losses than land-terminating glaciers. Up to 32% of the glacier mass loss is due to lake-terminating glaciers in this region, although these glaciers account for only 10% of the regional glacier population (King et al. 2019). Another 14% of ice loss is due to supraglacial lakes in this region, whose high energy absorption rate increases the melting of surrounding ice (Miles et al. 2018).

In addition to their impact on glacier melt, a large number of studies have focused on the potential hazard, which is deemed to be increasing due to the general growth in glacier lake volumes (Quincey et al. 2007; Fujita et al. 2013). Most multi-temporal lake inventories have used Landsat imagery as a baseline, with area-wide coverage continuously available since the late 1980s for most of the Himalaya (Zhang et al. 2015; Nie et al. 2017; Wang et al. 2020b; Chen et al. 2021; Zheng et al. 2021). According to the inventory by Nie et al. (2017), the number and area of glacier lakes have grown by 10% and 14.1%, respectively, between 1990 and 2015. Lakes in direct contact with their parent glaciers contributed 83.1% of the total lake increase. The area of supraglacial lakes more than quadrupled between 1990 and 2015, while the area of glacial lakes separated from their parent glacier hardly changed during this period (Nie et al. 2017).

Some of the highest regional growth rates prevail in the Central Himalaya, south of the main drainage divide (+23%). Many glaciers in this region are characterized by low flow velocities and low surface gradients in ablation areas, which favors rapid glacial ablation, formation of supraglacial pools and eventual coalescence into large lakes (Quincey et al. 2007; Benn et al. 2012). A handful of moraine-dammed lakes such as Imja Tsho, Lugge Tsho, South Lhonak, Zhangzangbo and Lower Barun (Figure 3.4(A) and (F)) have each grown by > 1 km^2 in recent decades, largely explaining the strong increase in lake area in this region (Haritashya et al. 2018; Wang et al. 2018; Maurer et al. 2019; Sattar et al. 2021). Negligible changes in the number and size of glacial lakes were observed in the northwestern Himalaya and Karakoram (Nie et al. 2017; Wang et al. 2020a). In contrast to the remainder of the Himalaya, glaciers in this region had an unchanged to positive mass balance and advanced to lower elevations (Farinotti et al. 2020), possibly limiting the formation of lakes to date.

Declassified Hexagon images have recently made it possible to extend the observation period for smaller study areas back to the 1960s and 1970s, thus revealing more about decadal trends in glacier lake changes. For example, the annual growth of lakes in the Nyainqentanglha Mountains (eastern

Himalaya) doubled from 0.04 km^2 yr^{-1} in 1976–2000 to 0.08 km^2 yr^{-1} in 2001–2018 (Luo et al. 2020). By contrast, lake growth decelerated in the Poiqu Basin (Central Himalaya), from 0.23 km^2 yr^{-1} between 1964 and 2010 to 0.03 km^2 yr^{-1} between 2010 and 2017 (Zhang et al. 2019). Similarly, a flatter trend was observed in the number of new lakes that formed in the Nepal, Central Himalaya, between 2007 and 2017, compared to the previous three decades (Khadka et al. 2018). Long-term increases in air temperatures, followed by glacier retreat, largely explain the long-term increase in lake abundance and size (Wang et al. 2017; Zhang et al. 2019). However, newly exposed ice-free areas are in higher and steeper terrain, which may have accounted for the reduced growth of glacial lakes in some regions in recent years (Zheng et al. 2021).

3.3.3. *Projections of future glacier lakes*

How will the distribution of glacier lakes change with future glacier retreat? The economic and ecological value of new lakes in mountains with decreasing ice cover might be in stark conflict with concomitant changes in hazard conditions. Lakes will gradually occupy suitable topographic depressions left by the glaciers and, in some cases, move closer towards steep mountain walls. These slopes can be prone to failure, as permafrost degradation reduces the internal stability of mountain walls (Krautblatter et al. 2013). Resulting landslides or rock avalanches can hit these newly forming water bodies and trigger flood waves that overflow the lake dams and turn into debris flows, endangering downstream valleys (Haeberli et al. 2017). Ongoing research therefore aims to predict where and when these new lakes will appear in the Himalaya (Linsbauer et al. 2016; Furian et al. 2021; Zheng et al. 2021). Such appraisals use estimates of the current glacier ice thickness, and project glacier retreat under different CO_2 emission scenarios until the Himalaya becomes largely ice-free eventually. Assuming that lakes completely fill these exposed depressions, $>$ 10,000 new glacial lakes could form with a total volume of 120 km^3 in an ice-free Himalaya (Linsbauer et al. 2016). Taking into account moderate (RCP2.6) and high (RCP8.5) emission scenarios, 56% or 86% of the additional lake area, respectively, will emerge until 2100 (Zheng et al. 2021). The most and largest lakes are projected to cluster in the Karakoram, where glaciers today cover the largest area. Regardless of the emission scenarios, the southern parts of the Himalaya will likely have attained the maximum exposure of lake area by 2050. In contrast, the Karakoram, drained by the

Indus River, will see a strong increase in lake area that is sustained into the late 21st century and beyond (Zheng et al. 2021). Thus, the region with the highest lake density may soon shift from the southwestern Himalayan region towards the Karakoram, possibly accompanied by a change in the GLOF hazard (Furian et al. 2021). More work is needed, however, to reduce substantial uncertainties in ice-thickness estimates (and thereby volume of the overdeepenings and associated future lake), as well as in the composition of dams (moraine or bedrock), both remaining elusive in such models.

3.4. Glacial lake outburst floods and downstream propagation

As illustrated above (Figure 3.4), glacial lakes are often ponded behind relatively weak dams made of unconsolidated rock, possibly interspersed with ice. For the water to be released, it usually takes a trigger that either causes the water to overtop the dam or the dam to lose its stability. Once this has happened, a chain reaction might evolve. The now flowing water will start eroding into the dam and create a breach which will eventually deepen and widen. The increased cross-section will enable flow velocities and discharge to increase even more, enhancing the flow's competence to entrain more dam material and even large boulders. The gushing waters will then propagate downstream. Close to the outburst lake, discharge will be much higher than during extreme meteorological floods (Schwanghart et al. 2016; Cook et al. 2018). However, with increasing flow distance, streams turn to rivers and these rivers become larger and less steep (Figure 3.5). This will prompt the outburst flood to deposit its coarse sediment load, and peak flow will attenuate. Triggering, breaching and propagation are the main processes that constitute the dynamics of a GLOF, and in the following sections, we will focus on these mechanisms.

3.4.1. *GLOF triggers*

Vuichard and Zimmermann (1987) vividly describes the scenery of the 1985 GLOF emanating from the lake Dig Tsho in the Khumbu Himal. "*The area was in full bloom and the potatoes were ready for harvesting. A colorful Sherpa festival was in progress when enormous amounts of water plummeted down the Dudh Kosi valley, destroying houses, bridges, trails, and much of the scarce agricultural land.*" Only a few people lost their lives but the floods destroyed infrastructure and killed livestock so that economic and social impacts were severe. What had happened?

Dig Tsho was a $\sim$0.5 km^2 lake at an elevation of 4,400 m a.s.l. in the uppermost Langmoche Valley. Situated above the lake, the Langmoche glacier had experienced rapid retreat beyond a rock step. August 4, 1985 was then a beautiful day with fine weather and intense radiation. The lake that had highly variable water levels was full of melt-water when an ice avalanche with a volume of 100,000–200,000 m^3 detached from the Langmoche glacier and plunged into the lake. This created a wave which traveled towards the lake outlet. As the lake narrows and the lake bottom rises towards the outlet, the wave increased its height to $\sim$5 m. With the lake being full to the brim, the wave easily overtopped the moraine dam and eroded it (Vuichard and Zimmermann 1987).

Many GLOFs have been initiated by waves caused by external triggers (Westoby et al. 2014). These triggers include glacier calving, ice and rock avalanches or landslides that impact on the lake surface so that their kinetic energy is dissipated through displacement waves (Roberts et al. 2014) or seiches which are standing waves that involve the entire water column (Westoby et al. 2014). The height and velocity of these waves are determined by the mass and velocity of the impacting ice and rock and modulated by lake bathymetry and form. If large enough, these waves may overtop and erode the dam. But even if they do not overtop the dam's crest, the waves may be eroding the dam structure and effectively weaken it. To this end, ice or rock avalanches as well as landslides may have their own triggers such as earthquakes or heavy rainfall which highlights that GLOFs may be an intermediate process in a rather complex hazard cascade.

Earthquakes, however, may also directly affect the stability of dams. As in artificial dams and reservoirs, ground shaking and fault movements may cause distortions of the dam by shearing, liquefaction and densification of the dam itself (Wieland 2014). Evidence for seismically triggered dam failure, however, is scarce (Strasser et al. 2008), and an analysis of the Gorkha Earthquake in 2015 revealed no discernible effects on GLOFs (Kargel et al. 2016).

Excessive input from rainfall and snow melt may lead to rapid lake filling, overflow and incision. The burst of Chorabari Lake upstream to the pilgrim city Kedarnath in Uttarakhand, for example, was triggered by heavy rainfalls during the India Floods 2013. Lack of a stable outflow led to a GLOF which added $\sim$400,000 m^3 water to the flood and exacerbated flood damages in Kedarnath (Allen et al. 2016).

Which of these triggers is responsible for a GLOF is often not easy to determine (Cook et al. 2018). If other precondition factors such as the height of the freeboard, dam geometry, and sedimentological or structural characteristics of the dam change, small triggers can have devastating consequences. For example, if the damming moraine is stabilized by an ice-core, climate warming may lead to a gradual degradation of the ice and structural weakening of the dam. To this end, the multitude of potential triggers as well as the role of preconditioning factors challenge our ability to reconstruct the causes of GLOFs and complicate the prediction of these events even more.

3.4.2. *Dam breach*

The wave that overtopped the moraine impounding Dig Thso eroded the downstream face of the dam, letting water spillover and form a breach channel. Now that flow started to drain the lake, it exerted shear stress on the walls of the initial breach. As moraines consist of a wide range of grain sizes, we can speculate that at the beginning, the escaping flow was only able to pick up the finer fraction forming the matrix of larger blocks. However, erosion may have been sufficient to widen and deepen the cut to instigate a positive feedback mechanism: down-cutting would increase discharge, increase shear stress and the capacity to erode, and thus the rate of down-cutting. The faster this process proceeds, the higher is the peak discharge which is a critical measure to characterize the hydrograph at the lake outlet (Vuichard and Zimmermann 1987).

There are several empirical, semi-physical and physically based, numerical models that can be used to simulate the breach-formation process (Westoby et al. 2014; Somos-Valenzuela et al. 2016). Unless the entire process cascade is encapsulated in a single simulation model (Worni et al. 2014; Mergili et al. 2017), hydrographs and peak discharge estimated from dam breach models are important to determine the boundary conditions of models that simulate downstream flood wave propagation (Schwanghart et al. 2016). A frequently used model to determine peak discharge is the physically based model of Walder and O'Connor (1997). Compared to empirical models that usually predict peak discharge from the lake volume, the model reveals the role of dam erosion in controlling peak discharges from overtopping dam failures.

The model by Walder and O'Connor (1997) is also able to account for the dam-break mechanism other than overtopping. For example, seepage and piping failures occur when water percolates through the dam and carries away the sediment. This may lead to a collapse of the dam crest above the pipe. Similar processes can be observed in ice-cored moraines when melting creates cavities in the dam that may collapse and cause slumping and dam-crest lowering.

3.4.3. *Flood propagation*

It took four to six hours for Dig Tsho lake to empty after the dam breach. Estimates of peak discharge, a metric characterizing a GLOF's impact and the ability to carry sediment, range between 6,000 and 8,000 m^3s^{-1} at the outlet of Dig Tsho. These high discharges were able to erode a total amount of 900,000 m^3 from the moraine including boulders with diameters of 2–5 m. The water and sediment mixture now traveled downstream, accompanied by enormous noise. The surge traveled fast at an average velocity of 4–5 $m.s^{-1}$ sustained for over 40 km. On its way, the GLOF dramatically modified the channel and valley morphology. In wide and low-gradient river sections, the river aggraded as soon as the surge front had passed, while the flow eroded and incorporated more material in narrow and steep reaches. The most extensive erosion and deposition occurred in the first 16 km of the GLOF route (Cenderelli and Wohl 2003). Further downstream, GLOF discharge and stream power attenuated, but were still able to destroy infrastructures such as a nearly completed hydroelectric power plant and roads.

Outburst floods exhibit three flow regimes. After a short phase of acceleration due to the hydrostatic reservoir pressure level, the flow quickly converges to an inertial regime characterized by high free-surface gradients, high flow velocities and turbulence. Further downstream, the flow then evolves to a viscous regime dominated by channel bed friction (Carrivick 2010). While flow in the first phase is mostly highly erosive, the inertial phase creates erosional and depositional forms, with the latter consisting of highly unsorted, mostly matrix-supported material derived from the natural dam. The momentum-dominated flow may further rework channel deposits and emplace spillover deposits where it leaves the channelized downslope path from the lake. Where sufficient material is available, the outburst flood will create a fan immediately downstream to the lake (Blair 2002). As soon as the flood enters the lower gradient sections of the river, it will attenuate and decrease stream power (see Volume 3 – Chapter 2). Channel modifications may still be profound, and patterns of sedimentation and erosion will vary with low and steep river gradients. However, with increasing distance downstream, the geomorphic effects will be reduced to further flood attenuation and because flood magnitudes will converge to those of extreme meteorological floods (Cenderelli and Wohl 2003, and Figure 3.5).

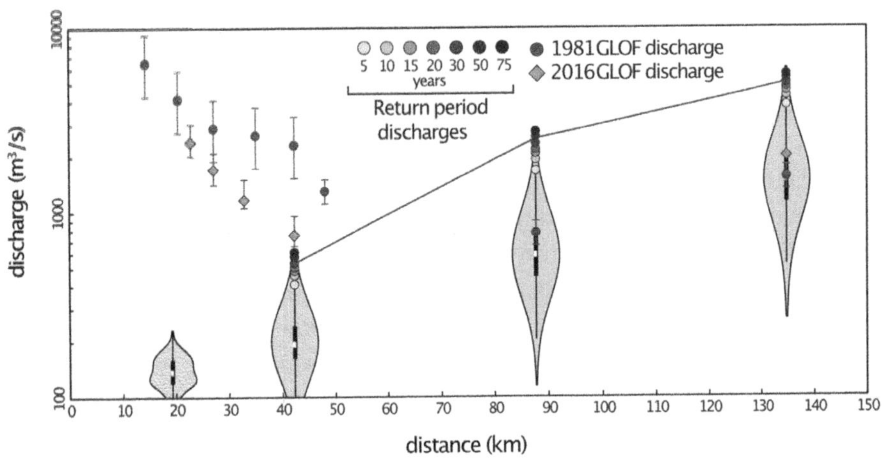

Figure 3.5. *Computed GLOF discharge versus monsoon discharge from Cook et al. (2018). Discharge with distance downstream of the Zhanzangbo confluence. Estimated discharges for the 1981 and 2016 GLOFs and the calculated discharges for floods or varying return periods (5–75 years). The blue line connects the 30-year return discharges. The violin plots show the full distribution of monsoon discharges (July and August) for each hydrological station (hydropower dam, Barabise, Parachuwatghat and Khurkot). For a color version of this figure, see www.iste.co.uk/cattin/himalaya3.zip*

Modeling and simulation of outburst floods is extremely challenging as floods modify their own channel bed, have very large variations in Reynold's numbers and vary strongly in terms of sediment concentrations, hence having solid and fluid phases involved that essentially affect flow dynamics (Pudasaini 2012). Simulating such flows can be achieved with software such as BASEMENT (Vanzo et al. 2021) or R.AVAFLOW (Mergili et al. 2017).

3.4.4. *Early warning*

Flow velocities exceeding 10 $m.s^{-1}$ make it very difficult to warn about GLOFs. A little lag time between alerts and arrival of the flood impedes any immediate measures to respond and to minimize the impact.

Notwithstanding, there are numerous efforts to install early warning systems that alert downstream communities once a GLOF occurs. Such systems consist of two components: a sensing system and a warning system. Sensing systems comprise manual observations or fully automated monitoring of either lake level (Kumar et al. 2020) or river discharge fluctuations (Gurung et al. 2010). More recently, Cook et al. (2021) demonstrated the potential of seismic networks to detect flash floods in real time (Figure 3.6) and provide early warning to downstream communities.

Issuing the alert is challenged by the difficulties of communication in remote mountainous terrain. Local communication systems have been deployed that create a field of Internet connectivity based on relay points along a line of sight (Gurung et al. 2010) while others employ satellite communication to transmit the data to an information hub that broadcasts the warning either via radio or activate air horns along the valley. Overall, harsh climatic conditions, inaccessibility, maintenance issues, security concerns and high operational costs render early warning systems very costly installations.

The system installed at Tsho Rolpa (Figure 3.7(A)), for example, is reported to involve costs of more than $1 Mio. (Gurung et al. 2010). Hence, manual warning systems are still common in both Nepal and Bhutan where such investments are rarely possible. Also, in the case of an outburst, instruments are often destroyed and washed away and cannot be refitted afterward (Figure 3.7(B)). Likewise, the siren systems are often vandalized and dismantled, but it is even more difficult to convey the information between the sensing devices and the siren in the rugged Himalayan landscape.

The experiences from the seismic monitoring during the Bhotekoshi flood (Cook et al. 2018, and Figure 3.7) and a more detailed feasibility investigation from the Chamoli flood, in Uttarakhand, India, on February 7, 2021 (Cook et al. 2021) underline the potential of seismic instruments for early warning. The Chamoli flood (Shugar et al. 2021) was very well detected by a dense array of seismometers (Cook et al. 2021), and exemplifies the potential of seismic networks to detect these types of catastrophic floods for potential early warning. This technology provides a new possibility to monitor catastrophic events with a few seismometers and might be even more needed in the future as it is expected that paraglacial extreme events occur more often in a warming climate.

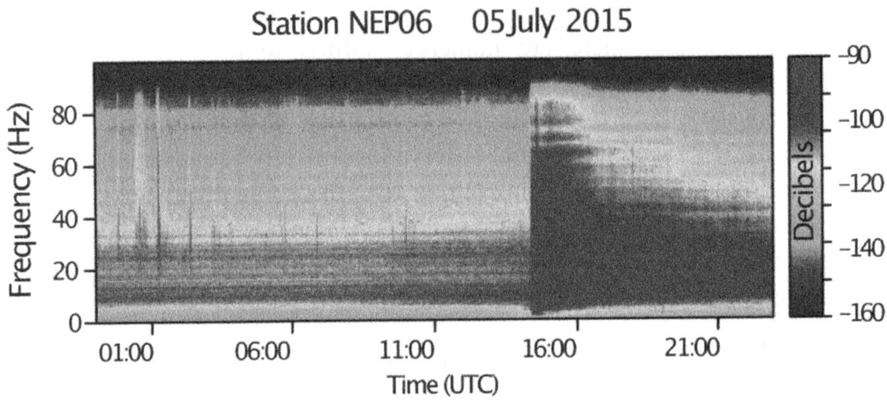

Figure 3.6. *Seismic spectrogram of the 2016 Bhotekoshi–Sunkoshi GLOF. The y-axis represents a frequency range of 0–100 Hz, and the x-axis shows time on July 5, 2016. The colors indicate the seismic power in each frequency band through time. Kindly provided by Kristen Lee Cook. For a color version of this figure, see www.iste.co.uk/cattin/himalaya3.zip*

Figure 3.7. *A) Climate monitoring systems in Tsho Rolpa glacial lake, Nepal. The monitoring is part of an early warning installation (photo: Sudan Bikash Maharjan, 2019). B) Destroyed pressure logger for water level monitoring by the 2016 Bhotekoshi–Sunkoshi GLOF (photo: Bhairab Sitaula). For a color version of this figure, see www.iste.co.uk/cattin/himalaya3.zip*

3.5. Consequences and impact

Depending on their discharge and sediment load, GLOFs may inflict severe damages tens to several hundreds of kilometers downstream. In the Himalaya, GLOFs have been recorded with discharges of 30,000 m^3s^{-1} and runout distances of about 200 km (Richardson and Reynolds 2000) and sedimentary evidence suggests that even larger floods occurred in the past (Schwanghart et al. 2016a; Srivastava et al. 2017). Such massive floods and their sudden arrival are particularly relevant to assess the risk of infrastructure, buildings and people nearby rivers draining glacierized catchments. In the context of climate change, not only the frequency of GLOF might increase in the future but also the nature of hazards can create a multi-hazard environment along with the changes in other cryospheric components.

There is a long list of destructive GLOF events reported in the Himalayan region. Ashraf et al. (2012) reported 35 events in the Karakoram range. At least 20 GLOF events recorded in the last seven decades resulted in heavy loss of human lives and property, agriculture, forest and destruction of infrastructure along the path of the floods. Similarly, between 2007 and 2008, five GLOF events had occurred in the Hunza basin of the Karakoram range which severely affected the downstream communities and infrastructure. Bajracharya et al. (2020) indicated about 50 GLOFs in the Hindu Kush Himalaya region, and most of the records are available only for some areas of China, Nepal, Pakistan and Bhutan. Nepal in the central Himalaya recorded 26 GLOF events with significant damage and loss of life. Subedi (2021) suggested that 40% of the GLOF events happened due to moraine dam failure in the region.

In Nepal, the eastern Koshi river basin is particularly exposed to GLOF events, with the transboundary Bhotekoshi–Sunkoshi GLOFs of 1964, 1981 and 2016 being prominent examples. While very little is known about the 1964 GLOF, the 1981 and 2016 GLOF events damaged the road network between China and Nepal and disturbed transportation for weeks (Cook et al. 2018; Subedi 2021). The 1981 GLOF had the most destructive effects in Nepal so far, with more than 200 people killed and total losses of about 3 million USD (ICIMOD 2011). The July 2016 GLOF originated from the small moraine-dammed Gongbatongshaco lake with a peak discharge up to 2,400 $m^3.s^{-1}$ at 30 km downstream from the lake (Figure 3.5). The GLOF event caused massive destruction, damaging 77 houses, three bridges and highways and destroying the intake dam of the 36 MW Upper Bhote Koshi Hydropower Project in Nepal (Liu et al. 2020, and Figure 3.8).

Figure 3.8. *Before and after photographs of the upper Bhotekoshi hydropower plant, located close to the Chinese border. A) Before the picture was taken in November 2006. The photograph shows the intake dam of the run by the river hydropower project. B) After the picture was taken in October 2016. The hydropower dam was partially destroyed and buried by the 2016 Bhotekoshi–Sunkoshi GLOF (photo: Christoff Andermann). For a color version of this figure, see www.iste.co.uk/cattin/himalaya3.zip*

Similarly to the Bhotekoshi–Sunkoshi events, the Dig Tsho GLOF of 1985 in the Dudh Koshi catchment (Mt Everest region) and the Sabai Tsho Tampokhari GLOF in 1998 caused a considerable loss of life, damaged infrastructure and affected livelihoods of people in the downstream areas (ICIMOD 2011). The Dig Tsho GLOF also destroyed the nearly completed Namche Small Hydroelectric project (ICIMOD 2011).

Khanal et al. (2015) carried out a comprehensive GLOF risk assessment for four glacial lakes located in Nepal and China under various outburst scenarios and their potential impacts on downstream infrastructure, people and property. The maximum affected level of GLOF damage potential of Thulagi (Gandaki basin, central Nepal), Lumuchimi China (north of Langtang), Tsho Rolpa (Upper Tama Koshi), Imja (Mt Everest region, Figure 3.1), located from west to east in Central Himalaya glacial is 36, 8, 415 and 188 million USD, respectively. The number of severely affected people range between 4,000 (Thulagi Lake) and 16,000 (Lumuchimi Lake). The maximum land affected would be in Imja, which is about 1,000 ha. The infrastructure affected by these GLOF scenarios includes schools, office buildings, temples, trails, roads, embankments, bridges, hydropower plants, irrigation canals, water mills and transmission lines. The indirect cost associated with the damage and disruption of these facilities is estimated much higher than the 8–415 million USD.

While there is only limited evidence that the frequency of GLOFs has changed (Hock et al. 2019), climate change is likely to exacerbate GLOF events (Ashraf et al. 2012) as rising temperatures particularly in the high altitudes cause accelerated melting, and expansion of existing glacial lakes. Moreover, GLOFs are considered to become more frequent as their triggers such as mass movements, avalanches and glacier collapses may become more often (Bolch et al. 2019; Vaidya et al. 2019).

Future GLOF risk will also be affected by the expansion of roads and power infrastructure into high-mountain areas (Fort et al. 2010). For example, Schwanghart et al. (2016) demonstrated that hydropower projects are now being planned and constructed in less than 40 km distance to glacierized areas and thus closer to glacial lakes. While tapping water resources along these reaches will benefit from high discharges and topographic gradients, GLOF magnitudes along these river reaches are increasingly difficult to assess due to the high uncertainties involved in outburst modeling.

3.6. Role in landscape formation

In the previous section, we described and documented the impacts of GLOFs on infrastructure and human lives. But these events also play a key role in modulating sediment fluxes, increasing rates of river incision and reworking sediments (see Volume 3 – Chapter 2). Thus, in the long run, GLOFs are likely to exert some control on the formation of the Himalayan mountain landscape. However, while the existence of outburst floods has been known for a long time, due to their rare nature and occurrence in sparsely populated high mountain areas, their role in mountain denudation and its quantification are still debated. Often, this discussion relied on the interpretation of remnants of such events. Pioneering work by Cenderelli and Wohl (2001) reconstructed peak discharges of the 1977 and 1985 GLOF events in the Dudh Koshi valley (Everest region) from floodmarks identified in photography and during field investigation. Based on these paleohydraulic reconstructions, they showed that GLOF magnitudes exceed those of meteorological floods by more than an order of magnitude in low-order rivers. Furthermore, Montgomery et al. (2004) investigated flood deposits in the Eastern Himalaya in and downstream of the Tsangpo Gorge and attributed these to very large moraine-dammed lakes storing $\sim$ 80 km^3 of water on the Tibetan side of the Himalayan range. Similarly, Scherler et al. (2014) linked sequences of fill and cut terraces in the Shyok Valley in the Karakoram to repeated moraine-dammed lake outburst floods. While these studies largely rely on the interpretation of field observations and numerical dating techniques, they are now increasingly supported by satellite-based remote sensing data and geophysical techniques as well as models that enable simulating such outburst events, thus providing additional evidence concerning their magnitudes.

Glaciers are known to erode effectively and provide large amounts of sediments but usually this was thought to be limited to the high elevations where glaciers can form. However, the downstream impact via associated effects such as GLOFs propagates the impact of glaciers downstream into the otherwise rain-dominated erosion domain. Dahlquist and West (2021) showed from topographic analysis that GLOFs effectively contribute to river incision, and thus erosion in the steep headwaters of the Himalaya. Furthermore, they show that outburst affected river valleys are generally narrower and steeper and they attribute this to the effectiveness of these high-magnitude, low-frequency events in scavenging sediments within the river channel in particular by mobilizing larger clasts. The removal of the bed amour in return exposes the bare bedrock to incision and promotes further incision of the river channel. In particular the mobilization of large boulders with $>$ 1 m diameters is rarely achieved by monsoon-driven floods. GLOFs and hyperconcentrated flows have the potential to transport boulders with meter to tens of meters diameter over long distances (Huber et al. 2020). Eventually, these boulders are emplaced in low-gradient river sections downstream and stabilize the river there and protect the channel bed against incision. Deposition from GLOFs or paraglacial processes can have tremendous impact further downstream and inundate the entire valleys in hundreds of meters of sediments, for example, around the two of Pokhara (Schwanghart et al. 2016).

The 2016 Bhotekoshi–Sunkoshi GLOF was one of the first detailed monitored events. It was detected by seismic instruments, situated away from the river capturing the seismic energy transmitted by the entire flood propagating downstream (Cook et al. 2018). All other instruments located in or close to the river were destroyed. However, the seismic record captured the flood as it propagated downstream. Bedload and river discharge inferred from the seismic record showed an enhanced bedload activity as well as suspended sediment load for roughly 14 days after the event. Changes of the river bed caused by the flood modified almost the entire river bed for ~40 km along the GLOF path. At several locations, incision of several meters was observed. The total geomorphic work done by the event exceeded any monsoonal flood. These findings illustrate the importance of GLOF events to mobilize large sediments and to incise the river bed. Furthermore, hillslope erosion triggered by lateral river migration and slope undercutting suggests that GLOFs impact landscapes beyond the fluvial system. The 2016 GLOF demonstrates the importance of quantifying the geomorphic processes of sediment mobilization and associated mass movements. One of the future research challenges will be to constrain the hitherto unquantified frequency of GLOFs which will be needed to quantify their relative role in shaping the Himalaya (Cook et al. 2018; Dahlquist and West 2021).

3.7. Conclusion

As in many mountain regions, glaciers in the Himalaya are mostly retreating. The "Karakoram Anomaly" (Hewitt 2005; Farinotti et al. 2020) stands out as a regional exception to this global trend, but as global warming continues, this trend is unlikely to proceed for many decades. Glacial retreat will affect Asia's water towers and thus likely water availability for irrigation and hydropower production, potentially putting food and energy security of tens of millions of people at risk. Moreover, glacial retreat and other cryospheric changes such as permafrost decay will modulate the frequency and magnitude of glacial hazards. GLOFs are a prime example of such hazards as they occur where glaciers retreat and leave water behind potentially unstable earthen dams that can suddenly breach. Once triggered, GLOFs can inflict damages tens to several hundreds of kilometers downstream from their source with devastating consequences for infrastructure, buildings and people, rendering their prediction and early warning necessary to secure mountain livelihoods. Owing to their erosiveness and capacity to transport large amounts of debris, GLOFs have a yet unquantified role in mountain denudation.

3.8. References

Allen, S.K., Rastner, P., Arora, M., Huggel, C., Stoffel, M. (2016). Lake outburst and debris flow disaster at Kedarnath, June 2013: Hydrometeorological triggering and topographic predisposition. *Landslides*, 13(6), 1479–1491.

Andermann, C., Longuevergne, L., Bonnet, S., Crave, A., Davy, P., Gloaguen, R. (2012). Impact of transient groundwater storage on the discharge of Himalayan rivers. *Nature Geoscience*, 5(2), 127–132.

Archer, D.R. and Fowler, H.J. (2004). Spatial and temporal variations in precipitation in the Upper Indus Basin, global teleconnections and hydrological implications. *Hydrology and Earth System Sciences*, 8(1), 47–61.

Ashraf, A., Naz, R., Roohi, R. (2012). Glacial lake outburst flood hazards in Hindukush, Karakoram and Himalayan ranges of Pakistan: Implications and risk analysis. *Geomatics, Natural Hazards and Risk*, 3(2), 113–132.

Azam, M.F., Ramanathan, A.L., Wagnon, P., Vincent, C., Linda, A., Berthier, E., Sharma, P., Mandal, A., Angchuk, T., Singh, V.B. et al. (2016). Meteorological conditions, seasonal and annual mass balances of Chhota Shigri Glacier, western Himalaya, India. *Annals of Glaciology*, 57(71), 328–338.

Azam, M.F., Wagnon, P., Berthier, E., Vincent, C., Fujita, K., Kargel, J.S. (2018). Review of the status and mass changes of Himalayan-Karakoram glaciers. *Journal of Glaciology*, 64(243), 61–74.

Azam, M.F., Kargel, J.S., Shea, J.M., Nepal, S., Haritashya, U.K., Srivastava, S., Maussion, F., Qazi, N., Chevallier, P., Dimri, A.P. et al. (2021). Glaciohydrology of the Himalaya-Karakoram. *Science*, 373(6557), eabf3668.

Bajracharya, Ratna, S., Bikash Maharjan, S., Shrestha, F., Chogyal Sherpa, T., Wagle, N., Bhakta Shrestha, A. (2020). Inventory of glacial lakes and identification of potentially dangerous glacial lakes in the Koshi, Gandaki, and Karnali river basins of Nepal, the Tibet Autonomous Region of China, and India. Lalitpur, Nepal [Online]. Available at: https://lib.icimod.org/record/34905.

Bazai, N.A., Cui, P., Carling, P.A., Wang, H., Hassan, J., Liu, D., Zhang, G., Jin, W. (2021). Increasing glacial lake outburst flood hazard in response to surge glaciers in the Karakoram. *Earth-Science Reviews*, 212, 103432.

Benn, D.I., Bolch, T., Hands, K., Gulley, J., Luckman, A., Nicholson, L.I., Quincey, D., Thompson, S., Toumi, R., Wiseman, S. (2012). Response of debris-covered glaciers in the Mount Everest region to recent warming, and implications for outburst flood hazards. *Earth-Science Reviews*, 114(1–2), 156–174.

Biemans, H., Siderius, C., Lutz, A.F., Nepal, S., Ahmad, B., Hassan, T., von Bloh, W., Wijngaard, R.R., Wester, P., Shrestha, A.B. et al. (2019). Importance of snow and glacier meltwater for agriculture on the Indo-Gangetic Plain. *Nature Sustainability*, 2(7), 594–601.

Blair, T.C. (2002). Alluvial-fan sedimentation from a glacial-outburst flood, Lone Pine, California, and contrasts with meteorological flood deposits. *International Association of Sedimentologists Special Publications*, 32, 113–140.

Bolch, T., Pieczonka, T., Mukherjee, K., Shea, J. (2017). Brief communication: Glaciers in the Hunza catchment (Karakoram) have been nearly in balance since the 1970s. *Cryosphere*, 11(1), 531–539.

Bolch, T., Shea, J.M., Liu, S., Azam, F.M., Gao, Y., Gruber, S., Immerzeel, W.W., Kulkarni, A., Li, H., Tahir, A.A. et al. (2019). Status and change of the cryosphere in the extended Hindu Kush Himalaya Region. In *The Hindu Kush Himalaya Assessment*, Wester, P., Mishra, A., Mukherji, A., Bhakta Shrestha, A. (eds). Springer International Publishing, Cham.

Bookhagen, B. and Burbank, D.W. (2010). Toward a complete Himalayan hydrological budget: Spatiotemporal distribution of snowmelt and rainfall and their impact on river discharge. *Journal of Geophysical Research*, 115(F3), F03019.

Brun, F., Berthier, E., Wagnon, P., Kääb, A., Treichler, D. (2017). A spatially resolved estimate of High Mountain Asia glacier mass balances from 2000 to 2016. *Nature Geoscience*, 10(9), 668–673.

Brun, F., Wagnon, P., Berthier, E., Jomelli, V., Maharjan, S.B., Shrestha, F., Kraaijenbrink, P.D.A. (2019). Heterogeneous influence of glacier morphology on the mass balance variability in High Mountain Asia. *Journal of Geophysical Research: Earth Surface*, 124(6), 1331–1345.

Carrivick, J.L. (2010). Dam break – Outburst flood propagation and transient hydraulics: A geosciences perspective. *Journal of Hydrology*, 380(3–4), 338–355.

Carrivick, J.L. and Tweed, F.S. (2013). Proglacial lakes: Character, behaviour and geological importance. *Quaternary Science Reviews*, 78, 34–52.

Carrivick, J.L. and Tweed, F.S. (2016). A global assessment of the societal impacts of glacier outburst floods. *Global and Planetary Change*, 144, 1–16.

Cenderelli, D.A. and Wohl, E.E. (2001). Peak discharge estimates of glacial-lake outburst floods and "normal" climatic floods in the Mount Everest region, Nepal. *Geomorphology*, 40(1–2), 57–90.

Cenderelli, D.A. and Wohl, E.E. (2003). Flow hydraulics and geomorphic effects of glacial-lake outburst floods in the Mount Everest region, Nepal. *Earth Surface Processes and Landforms*, 28(4), 385–407.

Chen, F., Zhang, M., Guo, H., Allen, S., Kargel, J.S., Haritashya, U.K., Watson, C.S. (2021). Annual 30 m dataset for glacial lakes in high mountain asia from 2008 to 2017. *Earth System Science Data*, 13(2), 741–766.

Cook, K.L., Andermann, C., Gimbert, F., Adhikari, B.R., Hovius, N. (2018). Glacial lake outburst floods as drivers of fluvial erosion in the Himalaya. *Science*, 362(6410), 53–57.

Cook, K.L., Rekapalli, R., Dietze, M., Pilz, M., Cesca, S., Rao, N.P., Srinagesh, D., Paul, H., Metz, M., Mandal, P. et al. (2021). Detection and potential early warning of catastrophic flow events with regional seismic networks. *Science*, 374(6563), 87–92.

Dahlquist, M. and West, A.J. (2021). Central Himalayan rivers record the topographic signature of erosion by glacial lake outburst floods. *Earth Surface Dynamics Discussions*, (August), 10, 705–722. https://doi.org/10.5194/esurf-10-705-2022.

De Kok, R.J., Kraaijenbrink, P.D.A., Tuinenburg, O.A., Bonekamp, P.N.J. (2020). Towards understanding the pattern of glacier mass balances in High Mountain Asia using regional climatic modelling. *The Cryosphere*, 14(9), 3215–3234. https://doi.org/10.5194/tc-14-3215-2020.

Farinotti, D., Round, V., Huss, M., Compagno, L., Zekollari, H. (2019). Large hydropower and water-storage potential in future glacier-free basins. *Nature*, 575(7782), 341–344.

Farinotti, D., Immerzeel, W.W., de Kok, R.J., Quincey, D.J., Dehecq, A. (2020). Manifestations and mechanisms of the Karakoram glacier Anomaly. *Nature Geoscience*, 13(1), 8–16.

Forsythe, N., Fowler, H.J., Li, X., Blenkinsop, S., Pritchard, D. (2017). Karakoram temperature and glacial melt driven by regional atmospheric circulation variability. *Nature Climate Change*, 7(9), 664–670.

Fort, M., Cossart, E., Arnaud-Fassetta, G. (2010). Hillslope-channel coupling in the Nepal Himalayas and threat to man-made structures: The middle Kali Gandaki valley. *Geomorphology*, 124(3–4), 178–199.

Fowler, H.J. and Archer, D.R. (2006). Conflicting signals of climatic change in the upper Indus Basin. *Journal of Climate*, 19(17), 4276–4293.

Fujita, K., Sakai, A., Nuimura, T., Yamaguchi, S., Sharma, R.R. (2009). Recent changes in Imja Glacial Lake and its damming moraine in the Nepal Himalaya revealed by in situ surveys and multi-temporal ASTER imagery. *Environmental Research Letters*, 4(4), 045205.

Fujita, K., Sakai, A., Takenaka, S., Nuimura, T., Surazakov, A.B., Sawagaki, T., Yamanokuchi, T. (2013). Potential flood volume of Himalayan glacial lakes. *Natural Hazards and Earth System Sciences*, 13(7), 1827–1839.

Furian, W., Loibl, D., Schneider, C. (2021). Future glacial lakes in High Mountain Asia: An inventory and assessment of hazard potential from surrounding slopes. *Journal of Glaciology*, 67(264), 653–670.

Gardelle, J., Arnaud, Y., Berthier, E. (2011). Contrasted evolution of glacial lakes along the Hindu Kush Himalaya mountain range between 1990 and 2009. *Global and Planetary Change*, 75(1–2), 47–55.

Ghimire, M. (2004). Review of studies on Glacier Lake outburst floods and associated vulnerability in the Himalayas. *Himalayan Review*, 35(0), 49–64.

Gurung, D.R., Bajracharya, S., Shrestha, B.R., Pradhan, P. (2010). Wi-Fi network at Imja Tsho (lake), Nepal. Glacial lake outburst flood (GLOF) Early Warning System (EWS) and rural connectivity. A dual significance. *Grazer Schriften Der Geographie Und Raumforschung*, 45, 321–326.

Haeberli, W., Schaub, Y., Huggel, C. (2017). Increasing risks related to landslides from degrading permafrost into new lakes in de-glaciating mountain ranges. *Geomorphology*, 293, 405–417.

Haritashya, U.K., Kargel, J.S., Shugar, D.H., Leonard, G.J., Strattman, K., Watson, C.S., Shean, D., Harrison, S., Mandli, K.T., Regmi, D. (2018). Evolution and controls of large glacial lakes in the Nepal Himalaya. *Remote Sensing*, 10(5), 1–31.

Herreid, S. and Pellicciotti, F. (2020). The state of rock debris covering Earth's glaciers. *Nature Geoscience*, 13(9), 621–627.

Hewitt, K. (2005). The Karakoram anomaly? Glacier expantion and the "elevation effect", Karakoram Himalaya. *Mountain Research and Development*, 25(4), 332–340.

Hock, R., Rasul, G., Adler, C., Cáceres, B., Gruber, S., Hirabayashi, Y., Jackson, M., Kääb, A., Kang, S., Kutuzov, S. et al. (2019). High mountain areas. In *Special Report on the Ocean and Cryosphere in a Changing Climate*, Pörtner, H.-O., Roberts, D.C., Masson-Delmotte, V., Zhai, P., Tignor, M., Poloczanska, E., Mintenbeck, K., Alegría, A., Nicolai, M., Okem, A. et al. (eds). Cambridge University Press, Cambridge and New York. https://doi.org/10.1017/9781009157964.004.

Huber, M.L., Lupker, M., Gallen, S.F., Christl, M., Gajurel, A.P. (2020). Timing of exotic, far-traveled boulder emplacement and paleo-outburst flooding in the central Himalayas. *Earth Surface Dynamics*, 8(3), 769–787.

Hugonnet, R., McNabb, R., Berthier, E., Menounos, B., Nuth, C., Girod, L., Farinotti, D., Huss, M., Dussaillant, I., Brun, F. et al. (2021). Accelerated global glacier mass loss in the early twenty-first century. *Nature*, 592(7856), 726–731.

ICIMOD (2011). Glacial lakes and glacial lake outburst floods in Nepal. ICIMOD, Kathmandu. doi:10.53055/ICIMOD.543.

Immerzeel, W.W. and Bierkens, M.F.P. (2012). Asia's water balance. *Nature Geoscience*, 5(12), 841–842.

Immerzeel, W.W., Lutz, A.F., Andrade, M., Bahl, A., Biemans, H., Bolch, T., Hyde, S., Brumby, S., Davies, B.J., Elmore, A.C. et al. (2020). Importance and vulnerability of the world's water towers. *Nature*, 577(7790), 364–369.

Iturrizaga, L. (2005a). Historical glacier-dammed lakes and outburst floods in the Karambar valley (Hindukush-Karakoram). *GeoJournal*, 63(1–4), 1–47.

Iturrizaga, L. (2005b). New observations on present and prehistorical glacier-dammed lakes in the Shimshal valley (Karakoram Mountains). *Journal of Asian Earth Sciences*, 25(4), 545–555.

Kargel, J.S., Leonard, G.J., Shugar, D.H., Haritashya, U.K., Bevington, A., Fielding, E.J., Fujita, K., Geertsema, M., Miles, E.S., Steiner, J. et al. (2016). Geomorphic and geologic controls of geohazards induced by Nepals 2015 Gorkha earthquake. *Science*, 351(6269), aac8353.

Khadka, N., Zhang, G., Thakuri, S. (2018). Glacial lakes in the Nepal Himalaya: Inventory and decadal dynamics (1977–2017). *Remote Sensing*, 10(12), 1–19.

Khanal, N.R., Mool, P.K., Shrestha, A.B., Rasul, G., Ghimire, P.K., Shrestha, R.B., Joshi, S.P. (2015). A comprehensive approach and methods for glacial lake outburst flood risk assessment, with examples from Nepal and the transboundary area. *International Journal of Water Resources Development*, 31(2), 219–237.

King, O., Bhattacharya, A., Bhambri, R., Bolch, T. (2019). Glacial lakes exacerbate Himalayan glacier mass loss. *Scientific Reports*, 9(1), 1–9.

Kok, R.J., Tuinenburg, O.A., Bonekamp, P.N.J., Immerzeel, W.W. (2018). Irrigation as a potential driver for anomalous glacier behavior in High Mountain Asia. *Geophysical Research Letters*, 45(4), 2047–2054.

Kraaijenbrink, P.D.A., Bierkens, M.F.P., Lutz, A.F., Immerzeel, W.W. (2017). Impact of a global temperature rise of 1.5 degrees Celsius on Asia's glaciers. *Nature*, 549(7671), 257–260.

Krautblatter, M., Funk, D., Günzel, F.K. (2013). Why permafrost rocks become unstable: A rock-ice-mechanical model in time and space. *Earth Surface Processes and Landforms*, 38(8), 876–887.

Kumar, B., Sathyan, A., Prabhu, T.S.M., Krishnan, A.K. (2020). Design architecture of glacier lake outburst flood (GLOF) early warning system using ultrasonic sensors. In *2020 IEEE Recent Advances in Intelligent Computational Systems (RAICS)*, 195–200. IEEE.

Lang, K.A., Huntington, K.W., Montgomery, D.R. (2013). Erosion of the Tsangpo Gorge by megafloods, Eastern Himalaya. *Geology*, 41(9), 1003–1006.

Linsbauer, A., Frey, H., Haeberli, W., Machguth, H., Azam, M.F., Allen, S. (2016). Modelling glacier-bed overdeepenings and possible future lakes for the glaciers in the Himalaya-Karakoram region. *Annals of Glaciology*, 57(71), 119–130.

Liu, M., Chen, N., Zhang, Y., Deng, M. (2020). Glacial lake inventory and lake outburst flood/debris flow hazard assessment after the Gorkha earthquake in the Bhote Koshi Basin. *Water (Switzerland)*, 12(2), 464.

Livingstone, S.J., Li, Y., Rutishauser, A., Sanderson, R.J., Winter, K., Mikucki, J.A., Björnsson, H., Bowling, J.S., Chu, W., Dow, C.F. et al. (2022). Subglacial lakes and their changing role in a warming climate. *Nature Reviews Earth and Environment*, 3(February), 106–124.

Luo, W., Zhang, G., Chen, W., Xu, F. (2020). Response of glacial lakes to glacier and climate changes in the western Nyainqentanglha range. *Science of the Total Environment*, 735, 139607.

Maharjan, S.B., Mool, P., Lizong, W., Xiao, G., Shrestha, F., Shrestha, R., Khanal, N., Bajracharya, S., Joshi, S., Shai, S. (2018). The status of glacial lakes in the Hindu Kush Himalaya-ICIMOD. Research report 2018/1.

Maurer, J.M., Schaefer, J.M., Rupper, S., Corley, A. (2019). Acceleration of ice loss across the Himalayas over the past 40 years. *Science Advances*, 5(6), eaav7266.

Mergili, M., Fischer, J.T., Krenn, J., Pudasaini, S.P. (2017). R.avaflow v1, an advanced open-source computational framework for the propagation and interaction of two-phase mass flows. *Geoscientific Model Development*, 10(2), 553–569.

Miles, E.S., Willis, I.C., Arnold, N.S., Steiner, J., Pellicciotti, F. (2017). Spatial, seasonal and interannual variability of supraglacial ponds in the Langtang Valley of Nepal, 1999–2013. *Journal of Glaciology*, 63(237), 88–105.

Miles, E.S., Willis, I., Buri, P., Steiner, J.F., Arnold, N.S., Pellicciotti, F. (2018). Surface pond energy absorption across four Himalayan Glaciers Accounts for 1/8 of total catchment ice loss. *Geophysical Research Letters*, 45(19), 10,464–10,473.

Montgomery, D.R., Hallet, B., Yuping, L., Finnegan, N., Anders, A., Gillespie, A., Greenberg, H.M. (2004). Evidence for Holocene megafloods down the Tsangpo River Gorge, Southeastern Tibet. *Quaternary Research*, 62(2), 201–207.

Mool, P.K., Wangda, D., Bajracharya, S.R. (2001). *Inventory of Glaciers, Glacial Lakes and Glacial Lake Outburst Floods: Monitoring and Early Warning Systems in the Hindu Kush-Himalayan Region – Bhutan*. International Centre for Integrated Mountain Development (ICIMOD), Kathmandu, Nepal.

Nie, Y., Sheng, Y., Liu, Q., Liu, L., Liu, S., Zhang, Y., Song, C. (2017). A regional-scale assessment of Himalayan glacial lake changes using satellite observations from 1990 to 2015. *Remote Sensing of Environment*, 189, 1–13.

Pfeffer, W.T., Arendt, A.A., Bliss, A., Bolch, T., Cogley, J.G., Gardner, A.S., Hagen, J.-O., Hock, R., Kaser, G., Kienholzm, C. et al. (2014). The Randolph glacier inventory: A globally complete inventory of glaciers. *Journal of Glaciology*, 60(221), 537–552.

Potocki, M., Mayewski, P.A., Matthews, T., Perry, L.B., Schwikowski, M., Tait, A.M., Korotkikh, E., Clifford, H., Kang, S., Sherpa, T. et al. (2022). Mt. Everest's highest glacier is a sentinel for accelerating ice loss. *NPJ Climate and Atmospheric Science*, 5, 7. https://doi.org/10.1038/s41612-022-00230-0.

Pritchard, H.D. (2019). Asia's shrinking glaciers protect large populations from drought stress. *Nature*, 569(7758), 649–654.

Pudasaini, S.P. (2012). A general two-phase debris flow model. *Journal of Geophysical Research: Earth Surface*, 117(3), 1–28.

Quincey, D.J., Richardson, S.D., Luckman, A., Lucas, R.M., Reynolds, J.M., Hambrey, M.J., Glasser, N.F. (2007). Early recognition of glacial lake hazards in the Himalaya using remote sensing datasets. *Global and Planetary Change*, 56(1–2), 137–152.

Richardson, S.D. and Reynolds, J.M. (2000). An overview of glacial hazards in the Himalayas. *Quaternary International*, 65–66, 31–47.

Rinzin, S., Zhang, G., Wangchuk, S. (2021). Glacial lake area change and potential outburst flood hazard assessment in the Bhutan Himalaya. *Frontiers in Earth Science*, 9(November), 1–25.

Roberts, N.J., McKillop, R., Hermanns, R.L., Clague, J.J., Oppikofer, T. (2014). Preliminary global catalogue of displacement waves from subaerial landslides. In *Landslide Science for a Safer Geoenvironment*, Sassa, K., Canuti, P., Yin, Y. (eds). Springer International Publishing, Cham.

Round, V., Leinss, S., Huss, M., Haemmig, C., Hajnsek, I. (2017). Surge dynamics and lake outbursts of Kyagar Glacier, Karakoram. *Cryosphere*, 11(2), 723–739.

Sattar, A., Goswami, A., Kulkarni, A.V., Emmer, A., Haritashya, U.K., Allen, S., Freyet, H., Huggel, C. (2021). Future glacial lake outburst flood (GLOF) hazard of the South Lhonak Lake, Sikkim Himalaya. *Geomorphology*, 388, 107783.

Scherler, D., Munack, H., Mey, J., Eugster, P., Wittmann, H., Codilean, A.T., Kubik, P., Strecker, M.R. (2014). Ice dams, outburst floods, and glacial incision at the western margin of the Tibetan Plateau: A > 100 k.y. chronology from the Shyok Valley, Karakoram. *Geological Society of America Bulletin*, 126(5–6), 738–758.

Scherler, D., Wulf, H., Gorelick, N. (2018). Global assessment of supraglacial debris-cover extents. *Geophysical Research Letters*, 45(21), 11,798–11,805.

Schwanghart, W., Bernhardt, A., Stolle, A., Hoelzmann, P., Adhikari, B.R., Andermann, C., Tofelde, S., Merchel, S., Rugel, G., Fort, M., Korup, O. (2016a). Repeated catastrophic valley infill following medieval earthquakes in the Nepal Himalaya. *Science*, 351(6269), 147–150.

Schwanghart, W., Worni, R., Huggel, C., Stoffel, M., Korup, O. (2016b). Uncertainty in the Himalayan energy–water nexus: Estimating regional exposure to glacial lake outburst floods. *Environmental Research Letters*, 11(7), 074005.

Schwanghart, W., Ryan, M., Korup, O. (2018). Topographic and seismic constraints on the vulnerability of Himalayan hydropower. *Geophysical Research Letters*, 45(17), 8985–8992.

Shean, D.E., Bhushan, S., Montesano, P., Rounce, D.R., Arendt, A., Osmanoglu, B. (2020). A systematic, regional assessment of high mountain asia glacier mass balance. *Frontiers in Earth Science*, 7(January), 1–19.

Shugar, D.H., Jacquemart, M., Shean, D., Bhushan, S., Upadhyay, K., Sattar, A., Schwanghart, W., McBride, S., Vries, M., Mergili, M. et al. (2021). A massive rock and ice avalanche caused the 2021 disaster at Chamoli, Indian Himalaya. *Science*, 373(6552), 300–306.

Somos-Valenzuela, M.A., Chisolm, R.E., Rivas, D.S., Portocarrero, C., McKinney, D.C. (2016). Modeling a glacial lake outburst flood process chain: The case of Lake Palcacocha and Huaraz, Peru. *Hydrology and Earth System Sciences*, 20(6), 2519–2543.

Srivastava, P., Kumar, A., Chaudhary, S., Meena, N., Sundriyal, Y.P., Rawat, S., Rana, N., Perumal, R.J., Bishat, P., Sharma, D. et al. (2017). Paleofloods records in Himalaya. *Geomorphology*, 284, 17–30.

Strasser, M., Schindler, C., Anselmetti, F.S. (2008). Late Pleistocene earthquake-triggered moraine dam failure and outburst of Lake Zurich, Switzerland. *Journal of Geophysical Research: Earth Surface*, 113(2), 1–16.

Subedi, S. (2021). Glacial Lake Outburst Flood (GLOF) hazard mitigation at Himalayan region, Nepal. PhD Thesis. University of East London.

Thompson, S.S., Benn, D.I., Dennis, K., Luckman, A. (2012). A rapidly growing moraine-dammed glacial lake on Ngozumpa Glacier, Nepal. *Geomorphology*, 145–146, 1–11.

Vaidya, R.A., Shrestha, M.S., Nasab, N., Gurung, D.R., Kozo, N., Pradhan, N.S., Wasson, R.J. (2019). Disaster risk reduction and building resilience in the Hindu Kush Himalaya. In *The Hindu Kush Himalaya Assessment*, Wester, P., Mishra, A., Mukherji, A., Bhakta Shrestha, A. (eds). Springer International Publishing, Cham.

Vanzo, D., Peter, S., Vonwiller, L., Bürgler, M., Weberndorfer, M., Siviglia, A., Conde, D., Vetsch, D. (2021). BASEMENT v3: A modular freeware for river process modelling over multiple computational backends. *Environmental Modelling and Software*, 143, 105102.

Veh, G., Korup, O., Walz, A. (2020). Hazard from Himalayan glacier lake outburst floods. *Proceedings of the National Academy of Sciences of the United States of America*, 117(2), 907–912.

Viviroli, D., Kummu, M., Meybeck, M., Kallio, M., Wada, Y. (2020). Increasing dependence of lowland populations on mountain water resources. *Nature Sustainability*, 3, 917–928.

Vuichard, D. and Zimmermann, M. (1987). The 1985 catastrophic drainage of a moraine-dammed lake, Khumbu Himal, Nepal: Cause and consequences. *Mountain Research and Development*, 7(2), 91.

Wagnon, P., Brun, F., Khadka, A., Berthier, E., Shrestha, D., Vincent, C., Arnaud, Y., Six, D., Dehecq, A., Ménégoz, M. et al. (2021). Reanalysing the 2007-19 glaciological mass-balance series of Mera Glacier, Nepal, Central Himalaya, using geodetic mass balance. *Journal of Glaciology*, 67(261), 117–125.

Walder, J.S. and O'Connor, J.E. (1997). Methods for predicting peak discharge of floods caused by failure of natural and constructed earthen dams. *Water Resources Research*, 33(10), 2337–2348.

Wang, S., Qin, D., Xiao, C. (2015). Moraine-dammed lake distribution and outburst flood risk in the Chinese Himalaya. *Journal of Glaciology*, 61(225), 115–126.

Wang, X., Chai, K., Liu, S., Wei, J., Jiang, Z., Liu, Q. (2017). Changes of glaciers and glacial lakes implying corridor-barrier effects and climate change in the Hengduan Shan, southeastern Tibetan Plateau. *Journal of Glaciology*, 63(239), 535–542.

Wang, W., Gao, Y., Anacona, P.I., Lei, Y., Xiang, Y., Zhang, G., Li, S., Lu, A. (2018). Integrated hazard assessment of Cirenmaco glacial lake in Zhangzangbo valley, Central Himalayas. *Geomorphology*, 306, 292–305.

Wang, X., Guo, X., Yang, C., Liu, Q., Wei, J., Zhang, Y., Liu, S., Zhang, Y., Jiang, Z., Tang, Z. (2020a). Glacial lake inventory of high-mountain Asia in 1990 and 2018 derived from Landsat images. *Earth System Science Data*, 12(3), 2169–2182.

Wang, X., Guo, X., Yang, C., Liu, Q., Wei, J., Zhang, Y., Liu, S., Zhang, Y., Jiang, Z., Tang, Z. (2020b). Glacial lake inventory of High Mountain Asia (1990–2018) derived from Landsat images. *Earth System Science Data Discussions*, 12, 1–23.

Watson, C.S., King, O., Miles, E.S., Quincey, D.J. (2018). Optimising NDWI supraglacial pond classification on Himalayan debris-covered glaciers. *Remote Sensing of Environment*, 217(August), 414–425.

Wendleder, A., Schmitt, A., Erbertseder, T., D'Angelo, P., Mayer, C., Braun, M.H. (2021). Seasonal evolution of supraglacial lakes on Baltoro Glacier from 2016 to 2020. *Frontiers in Earth Science*, 9(December), 1–16.

Westoby, M.J., Glasser, N.F., Brasington, J., Hambrey, M.J., Quincey, D.J., Reynolds, J.M. (2014). Modelling outburst floods from moraine-dammed glacial lakes. *Earth-Science Reviews*, 134, 137–159.

Wieland, M. (2014). Seismic hazard and seismic design and safety aspects of large dam projects. In *Perspectives on European Earthquake Engineering and Seismology. Geotechnical, Geological and Earthquake Engineering*, Ansal, A. (ed.). Springer, Cham. https://doi.org/10.1007/978-3-319-07118-3_20.

Worni, R., Huggel, C., Clague, J.J., Schaub, Y., Stoffel, M. (2014). Coupling glacial lake impact, dam breach, and flood processes: A modeling perspective. *Geomorphology*, 224, 161–176.

Zhang, G., Yao, T., Xie, H., Wang, W., Yang, W. (2015). An inventory of glacial lakes in the Third Pole region and their changes in response to global warming. *Global and Planetary Change*, 131, 148–157.

Zhang, G., Bolch, T., Allen, S., Linsbauer, A., Chen, W., Wang, W. (2019). Glacial lake evolution and glacier-lake interactions in the Poiqu River basin, central Himalaya, 1964–2017. *Journal of Glaciology*, 65(251), 347–365.

Zheng, G., Allen, S.K., Bao, A., Ballesteros-Cánovas, J.A., Huss, M., Zhang, G., Li, J., Yuan, Y., Jiang, L., Yu, T. et al. (2021). Increasing risk of glacial lake outburst floods from future Third Pole deglaciation. *Nature Climate Change*, 11(5), 411–417.

4

Landsliding in the Himalaya: Causes and Consequences

Odin MARC[1], Kaushal GNYAWALI[2,3], Wolfgang SCHWANGHART[4] and Monique FORT[5]

[1] *Paul Sabatier University, Toulouse, France*
[2] *School of Engineering, University of British Columbia, Kelowna, Canada*
[3] *Natural Hazards Section, Himalayan Risk Research Institute, Bhaktapur, Nepal*
[4] *University of Potsdam, Germany*
[5] *Paris Cité University, France*

4.1. Introduction

Landsliding is a major hazard in active mountain ranges, as well as a key agent of long-term erosion (see Volume 3 – Chapter 2). The Himalayan range is very susceptible to landsliding due to its steep topography, frequent large magnitude earthquakes and its strong seasonal cycle with the Indian summer monsoon pouring heavy rainfall continuously during three months, every year. Beyond frequent landsliding caused by the hydrometeorological cyclic forcing, the Himalaya are also exposed to a high seismicity which results in several large shallow earthquakes per century, some likely among

Himalaya, Dynamics of a Giant 3,
coordinated by Rodolphe CATTIN and Jean-Luc EPARD.

the largest continental earthquakes recorded (see Volume 1 – Chapter 4 and Volume 3 – Chapter 5), and recurrent catastrophic glacial lake outburst floods (see Volume 3 – Chapter 3). Finally, the exceptional topography of the range, hosting the Earth's highest peaks and deepest gorges, is likely a condition for massive failure, of whole peaks or ridges, leaving among the largest deposits on Earth (>10 km^3).

The significance of these multiple causes of landslides is outstanding given that the Himalaya are populated by >100 millions inhabitants, and host a dense network of mountainous communities, ongoing road constructions and a large number of operational or planned hydroelectric facilities. These human activities are often vulnerable to landslide hazard and to some extent may even enhance slope instabilities.

In the first part of this chapter, we review the state-of-the-art and current challenges concerning our understanding of the natural controls of landsliding in the Himalaya. In the second part, we explore the links between landsliding and societies in the Himalaya as well as potential for mitigation, in the context of global climate change.

4.2. Understanding landsliding and their links to the dynamics of the Himalayan range

4.2.1. *Preliminary notions on the mechanics of landsliding*

Landslides occur at a variety of scales, from small shallow (< 1–3 m) soil failures to large deep-seated bedrock landslides (Larsen et al. 2010). Nevertheless, a simple approach to study slope stability is to consider a landslide failure envelope, on which a force balance can be made, and a safety factor defined by the ratio of resisting to driving forces (De Blasio 2011; Milledge et al. 2014).

In static, dry conditions, the safety factor is the ratio of frictional and cohesive forces to the (projected) weight of the rockmass above the failure envelope. The key effect of rainfall is to rise the pore pressure level above the failure envelope and thus to reduce the frictional resistance (as the effective normal load is reduced). The increase in pore pressure is commonly due to both rising water table over a long period and/or transient pulse of rainfall infiltration during intense downpour (Van Asch et al. 1999; Iverson 2000). Whether dry or wet, a slope can experience strong motion during an earthquake that can be modeled as an additional acceleration which will alter the stress in both directions, normal and parallel to the failure (Newmark 1965).

Strong motion can also shatter and deform rockmass and soil, causing coseismic landslides (Meunier et al. 2007, 2013; Marc et al. 2016), as well as causing ground cracking which may enhance failure propensity during the following rainfall events, leading to the so-called post-seismic landsliding (Marc et al. 2015, 2021, see Figure 4.1).

Finally, beyond these two most common triggers, other processes may change the force balance on a hillslope, leading to landslide movement and failure, such as slope debuttressing due to glacier retreat or erosion at the base of a slope, either due to incision and lateral migration of rivers or due to road construction.

When a hillslope section becomes unstable, it may fail rapidly and completely or move slowly, deforming the ground and bending trees. The former, here called rapid landslides, are the most common and occur at all sizes in all materials, while the latter, called slow-moving or creeping landslides, are more often rather large and deep landslides, and often occur in weak bedrock. Still, it is important to remember that slow-moving landslides may fail completely at some point, usually after a phase of progressive acceleration of the displacement, likely associated with a weakening of the basal plane (see Petley et al. 2005; Handwerger et al. 2016), and thus it is likely that a number of large deep-seated landslides with rapid failure may have been preceded by slow deformation of at least part of the slope (e.g. Chigira et al., 2013). A recent example of a slow-moving landslide transitioning into a catastrophic and rapid slope failure is the Jure (Sunkoshi) landslide which occurred in 2014 in Nepal. Here, pre-event optical and radar imagery revealed repeated modest landsliding (Marc et al. 2019) and seasonal downslope displacement (Ao et al. 2020) on the upper part of the future failure. These different triggering and deformation mechanisms will now be related to the Himalayan climatic and tectonic setting.

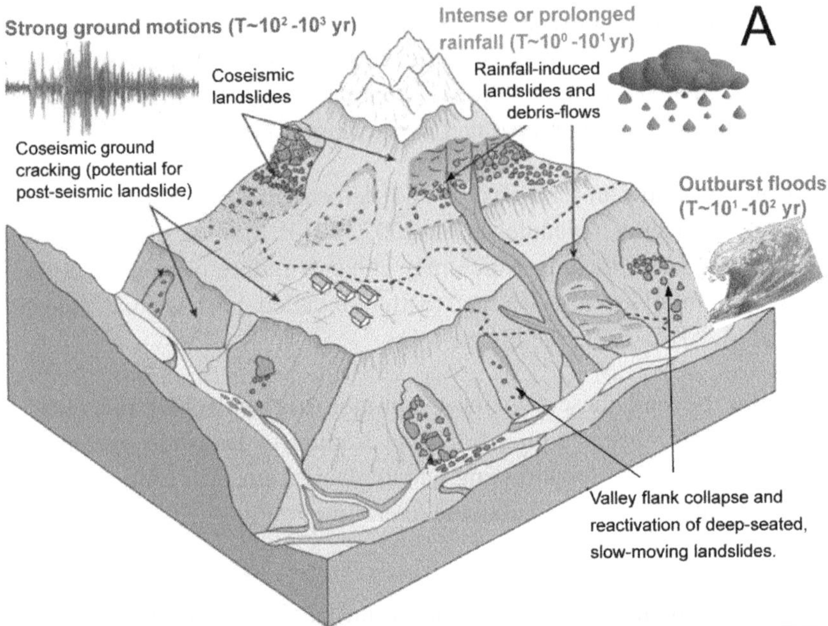

Figure 4.1. *A: Illustration of recurrent triggering events in the Himalaya (with their typical return time) and the associated landsliding. Modified from Rosser et al. (2021). B: View of the Bhote Koshi valley in June 2015 illustrating the diversity of landslide size, type and location on the hillslope, most of them caused by the Gorkha earthquake (photo courtesy, Kristen Cook, GFZ). For a color version of this figure, see www.iste.co.uk/cattin/himalaya3.zip*

4.2.2. Seasonal landsliding caused by monsoons and extreme rainfall

Throughout the Himalaya, landslides occur predominantly during the monsoon season, which is approximately from June to late September. Based on a selection of precisely dated landslides and local rain gauges, several studies have defined rainfall thresholds for shallow landsliding across the Himalaya, in Uttarakhand, Nepal, Bhutan or Darjeeling (e.g. Dahal and Hasegawa 2008; Mathew et al. 2014; Dikshit et al. 2019). Such studies are essential to understanding landslide hazards and the design of early warning systems, but are often limited by the small number of studied landslides, sparse constraints on rainfall variability in the study area, and the difficulty to account for other controls on landslide triggering which include local ground properties and the hydrological conditions that arise from antecedent rainfall (Bogaard and Greco 2016). Using suspended load time series as a proxy for landslide occurrence, Gabet et al. (2004) found that landslides in the Annapurna region occurred only after a minimum accumulated monsoon rainfall, interpreted as the required water volume to reach a regolith moisture level allowing gravity drainage, and required a minimum daily rainfall amount decreasing through the monsoon. They attributed the variability in landslide initiation to variable slope gradient and regolith storage (i.e. thickness and porosity). There was limited progress in developing and constraining such physical model, and new insight would require more systematic constraints on when landslides happen during the monsoons (e.g. Burrows et al., 2022).

More recently, several studies have produced comprehensive landslide inventories in parts of the Himalayan range, attempting to link monsoon-induced landslides to hydrometeorologic variables. In four valleys of central Nepal, hundreds of landslides were mapped for each monsoon between 2010 and 2017, but a consistent correlation with various monsoonal metrics, constrained by a recalibrated satellite measurement of rainfall, could not be found (Marc et al. 2019).

In contrast, in a large catchment of Western Nepal, after mapping and dating (at the annual scale) a comprehensive landslide inventories from 1992 to 2018, a good correlation ($R \sim 0.7$) emerged between the total shallow

(< 10 m) landslide number and ERA5 reanalysis climatic variables such as the total precipitation or maximum precipitation over 10 days during the monsoon (Munoz-Torrero Manchado et al. 2021). This correlation was most pronounced in the High Himalaya, and was completely absent for deep-seated landslides, consistent with the observations that large landslide could depart from the power-law scaling statistics observed for a given monsoon (Marc et al. 2019). This suggests that such deep-seated failures likely relate rather to long-term slope deformation and rockmass fatigue (Lacroix and Amitrano 2013; Handwerger et al. 2016) rather than monsoon-driven pore pressure increases. Interestingly, this study also suggested that landslide number may be moderately anticorrelated to the previous monsoon total rainfall, so that landsliding in a given year would not only depend on the current year rainfall but also the previous one. Similarly, the comprehensive mapping of annually resolved monsoon landsliding in Central and Eastern Nepal revealed a strong exponential increase of total landslide volume with total monsoon rainfall, either from PERSIANN satellite product or APHRODITE raingauge interpolated product (Jones et al. 2021, see Figure 4.2A).

This recent study highlights the strength of comprehensive, long-term mapping of landslide, which, even when done with low-resolution imagery, allowed us to average out the spatio-temporal variability of confounding factors and reveal trends with the hydroclimatic drivers. This study also showed that anomalous years with exceptional landsliding relative to the scaling were related to intense localized rainfall within these years which had caused a local burst of landsliding (Bell et al. 2021; Jones et al. 2021, see Figure 4.2A). Indeed, localized intense rainfall is regularly hitting small areas of the Himalaya, often causing landslides and debris flow with catastrophic consequences (Hobley et al. 2012; Dimri et al. 2017; Kumar et al. 2017, and references therein). However, most of these studies focus on few landslides associated with these cloudbursts. To our knowledge, Jones et al. (2021) performed the first study which could clearly highlight the superposition of background monsoonal triggering and exceptional, localized forcing on top of it. A hint of the importance of localized intense rainfall on seasonal landsliding could also be inferred from the observation of landslide cluster moving within the catchment from one year to another (Marc et al. 2019).

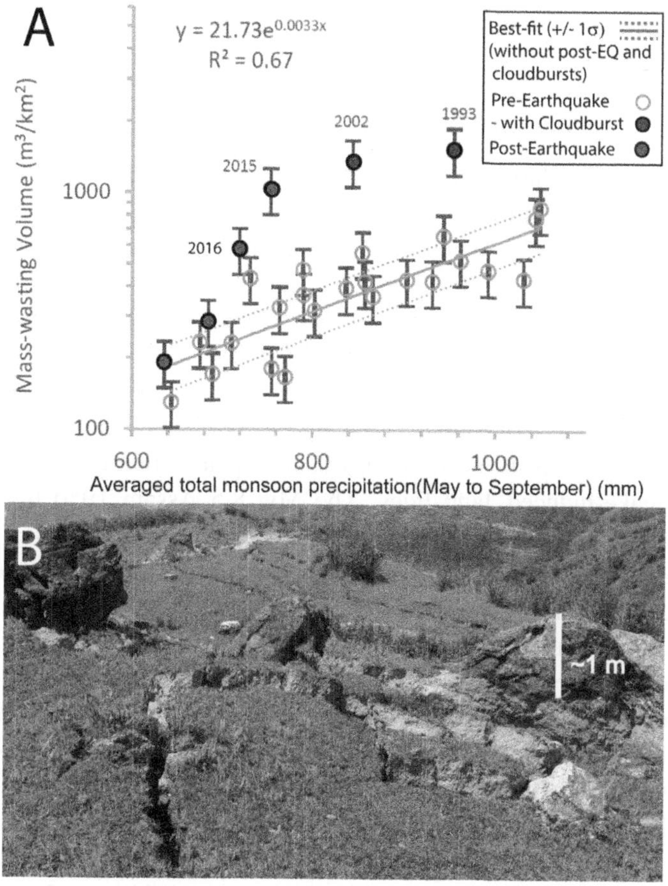

Figure 4.2. *A: Total landslide volume derived from scars of new and remobilized landslides mapped with Landsat imagery against the total monsoon precipitation in an area of Nepal of 42,000 km^2, affected by the Gorkha earthquake, from 1988 to 2018 (when imagery was available). Rainfall was estimated from a long-term satellite-based product (PERSIANN-CDR, Ashouri et al. 2015), but similar results could be obtained with gauge products (i.e., APHRODITE, see Yatagai et al. 2012). Note how the first monsoons following the Gorkha earthquake as well as the monsoons during which extreme daily precipitation occurred cause exceptional landsliding relative to the trend defined by the other years. Modified from Jones et al. (2021). B: Picture of a ridge in the Bhote Koshi valley in June 2015 showing meter deep ground cracks, often longer than 10 meters, formed by the Gorkha earthquake and possibly favoring post-seismic landsliding (photo courtesy from A. Schoepa, GFZ-Potsdam). For a color version of this figure, see www.iste.co.uk/cattin/himalaya3.zip*

4.2.3. *Landslide induced by earthquakes and other exceptional perturbations*

4.2.3.1. *Coseismic landsliding*

Earthquakes of large magnitude (>7) are very frequent in the Himalaya, with an estimate return time of a few hundred years for 7.5<Mw<8.5 (Bollinger et al. 2014; Avouac 2015). Stevens and Avouac (2016) conclude that even earthquakes with Mw>9.0 occur at a return period of 800 years (see Volume 3 – Chapters 5 and 6). Based on the global database analysis and given the steep slopes prevailing throughout the Himalaya, such large magnitude earthquakes are expected to cause thousands to tens of thousands of landslides over large areas, typically 1,000–30,000 km^2 (Keefer 1984; Marc et al. 2017). Still until the 2000s, few quantitative data were available on earthquake-induced landslides. The Mw 8.5 Assam 1950 had triggered widespread landsliding throughout northeast India, and was holding the record for the largest single erosional event associated with an earthquake, with an estimated total landslide volume of 47 km^3 (Mathur 1953; Keefer 1994). However, this value is very uncertain, as the extent of landsliding was only based on field surveys, and as the average failure depth was used to convert this overall area into depth (Mathur 1953). Then, in the 2000s, several earthquakes occurred during the era of high-resolution satellite observations, namely, the Mw 7.5 Kashmir 2005, the Mw 6.9 Sikkim 2011 and the Mw 7.8 Gorkha 2015, allowing comprehensive mapping of the coseismically triggered landslides and an analysis of their pattern and implications. The Mw 7.5 Kashmir earthquake triggered relatively few landslides (<5,000), mainly rockfall and debris fall including the large Hattian Bala rock avalanche (Table 4.1, Owen et al. 2008). The landslides concentrated along the ruptured fault (< 5 km), on very steep slopes (> 30°) and specific lithological formations. Additionally, 53% of landslides were associated with road construction (Owen et al. 2008), further suggesting that the earthquake mainly triggered sites with strong predisposition towards failure. Similarly, the Mw 6.8 Sikkim earthquake triggered a few hundred shallow landslides mainly on highly susceptible sites such as road cuts, terrace deposits and near the crest of sharp ridges close to the epicenter (Chakraborty et al. 2014).

Location	Catchment	N	E	Landslide type	Rock type	Tectonic unit	Estimated volume ($10^9\ m^3$)	Age	Ref.
Doian	Astor, Pakistan	35.52	74.68	RS / DS	GN	HHC	5.4	Prehistoric, 18th/19th century	Shroder Jr. (1998)
Mushkin	Astor, Pakistan	35.49	74.73	RS / DS	GN	HHC	8.2	Prehistoric, 18th/19th century	Shroder Jr. (1998)
Hattian	Azad Kashmir, N Pakistan	34.08	73.43	RS / DS	SE	SI	0.08	2005 Kashmir earthquake	Owen et al. (2008)
Keylong Serai	India	32.48	77.26	RA	MS	TTH	0.9	7510±110 (^{10}Be age)	Mitchell et al. (2007)
Sichling	Spiti, India	32.11	78.18	RS	SE	TTH	1.4	8751 BP (^{14}C age)	Bookhagen et al. (2005)
Rekong Peo	Sutlej, India	31.52	78.25	RS	PG	HHC	1	Mid-Holoce	Bookhagen et al. (2005)
Chango	Spiti, India	32.07	78.59	RS	GN	TTH	1	28.92	Bookhagen et al. (2005)

Table 4.1. *Selected characteristics of giant ($> 10^8\ m^3$) catastrophic landslides in Pakistan and India, modified and completed from Weidinger and Korup (2009). Landslide types: RA: rock avalanche; RS: rockslide; DS: debris slide. Rock type: SE: Sedimentary; MS: Metasedimentary; PG: Paragneiss; GN: Gneiss. Tectonic unit: HHC: Higher Himalayan Crystalline complex; TTH: Tibetan (Tethyan) metasediments; SI: SIwaliks*

The Mw 7.8 Gorkha earthquake was followed by a rapid mapping effort done by multiple researchers (Kargel et al. 2016; Martha et al. 2017; Gnyawali and Adhikari 2017; Roback et al. 2018; Valagussa et al. 2021, and Volume 3 – Chapter 8) with the most comprehensive inventory containing more than 25,000 landslides (Roback et al. 2018). The largest aftershock, the Mw 7.3 Dolakha earthquake, also caused several hundreds of landslides in the Bhote Koshi area (Marc et al. 2019, see Figure 4.1B) and the Dolakha district (Martha et al. 2017). Landslides focused on steep slopes but occurred throughout all lithological units of the High Himalaya, consistent with the strong shaking in this area and moderate rock strength (Gallen et al. 2017).

In the Siwaliks range, which displays abundant seasonal landsliding (Jones et al. 2021), there was almost no coseismic landsliding (Roback et al. 2018) which can only be explained with moderately strong rock strength (Gallen et al. 2017), somewhat surprising for the regions, made of rapidly uplifting detrictic sedimentary layers.

At regional scale, the coseismic landslide pattern clearly bears the mark of the eastward rupture directivity with the landslide zone extending far to the east of the fault plane and the landslide occurring significantly more frequently in the hillslope facing eastward and southward (Gnyawali and Adhikari 2017; Tian et al. 2020).

For various large earthquakes with available ground motion measurements, landslide density was shown to be highly correlated to the peak ground acceleration (PGA) (Meunier et al. 2007, 2013; Yuan et al. 2013). However, given that only Shakemaps PGA estimates are available in Nepal, mainly based on empirical models calibrated by teleseismic and felt reports, the correlation between landslide density and ground acceleration (PGA) remains moderate (Martha et al. 2017; Roback et al. 2018). Thus, research focusing on using better seismological models, such as 3D models including kinematic fault rupture, or topographic amplification is still needed (e.g. Dunham et al., 2022).

4.2.3.2. *Post-seismic landsliding*

Elevated landslide response to rainfall following several recent earthquakes suggested that the strong shaking could damage the near surface and lead to "post-seismic" landsliding (see Marc et al. 2015, 2021). In the case of the Kashmir earthquake in 2005, following the observations of widespread coseismic cracking (Owen et al. 2008), several zones were monitored for several years, with some zones showing no evidence of exceptional landsliding (Khattak et al. 2010) and others showing a peak of landsliding in 2006 (Saba et al. 2010). However, given this was associated with exceptional monsoon rainfall (Saba et al. 2010), it is not clear if landslide susceptibility to rainfall had been elevated by the earthquake or not. In contrast, after the Gorkha earthquake, widespread cracking was also observed (see Figure 4.2(B)); several catchments displayed 5–10 times more landsliding in 2015 than the 2010–2014 average monsoon landsliding, while the monsoon was rather weaker than in the previous years (Marc et al. 2019). In 2016 and 2017, the rate of new landsliding was decaying and back to normal, respectively. Geophysical monitoring of the subsurface, through ambient seismic noise correlation, also suggests a rapid recovery of the near-surface properties, although the link with landsliding recovery is better supported for other earthquakes (Marc et al. 2021).

These observations are consistent with an elevated rate of new debris flow occurrence in 2015 within the area affected by coseismic landsliding, rapidly decaying in the following years (Dahlquist and West 2019). The recent quantitative scaling relationship between monsoon properties and seasonal landsliding constrained by Jones et al. (2021, see Figure 4.2A) over most of the rupture area has confirmed and strengthened these preliminary estimates. Nevertheless, further research is required to constrain the processes controlling the timescales and magnitude of the susceptibility changes, probably through innovative geophysical monitoring in the field (e.g. Marc et al. 2021). Also note that landslide and debris-flow susceptibility may transiently rise after other events such as large wildfire (e.g. DeGraff et al. 2015), which may occur in the Himalaya (Parajuli et al. 2020), and become more frequent with the ongoing climate change.

4.2.3.3. *Landsliding associated with outburst floods*

Outburst flood is another type of episodic perturbation, defined by a high and transient river discharge typically due to the rapid injection of large water volume into the river network (see Volume 3 – Chapter 4), which can lead to widespread landsliding on the valley flanks (Larsen and Montgomery 2012; Cook et al. 2018, see Figure 4.1A). Outburst flood may originate from large landslide dam failure, as in the Zhamu creek (Tibet) in 2000 (Larsen and Montgomery 2012), or because of the breaching or overspilling of a glacial lake, as in the Bhote Koshi (Nepal) in 2016 (Cook et al. 2018) or Makalu (Nepal) 2017 (Byers et al. 2019).

Specifically, many recent landslides in the Tsangpo Gorge (Tibet) were located on the outer bend of meanders, suggesting that their failure was due to intense lateral erosion during the Zhamu Creek outburst flood (Larsen and Montgomery 2012). The contribution of outburst floods to long-term erosion by landsliding is a current research challenge, but it is clear that they substantially impact the erosional dynamics of the valley flanks. Indeed, attributing all the bank collapse in the Bhote Koshi (Nepal) to the 2016 glacial lake outburst floods, the erosion rate on the $\sim$25 km long corridor along the main river trunk, considering only the lower half of the slope, was two orders of magnitude larger than the mean 2010–2014 erosion in the same corridor and equivalent to about five times the mean 2010–2014 total monsoon landslide volume in the surrounding $\sim$1,000 km^2 (Cook et al. 2018; Marc et al. 2019). However, a key challenge remains to resolve the frequency of such floods (e.g. Veh et al. 2019) and to better characterize the spatial pattern of flood-induced landsliding alongstream, as the flood discharge becomes less and less exceptional while traveling downstream.

4.2.4. *Giant and paleo-landslides*

Landslide volumes measured over large areas usually follow long-tailed distributions. This means that most landslides have relatively small sizes, but that a few can have very large volumes (see Volume 3 – Chapter 2 for "power law distribution" in geomorphology). Landform and sedimentological evidences throughout the Himalaya demonstrate that large to giant landslide deposits (i.e. $> 10^6$ to $> 10^8$ m^3) (Korup et al. 2006, 2007, see Figure 4.3, Tables 4.1 and 4.2) are not rare in the Himalaya, even if many of them may still be undetected.

Indeed, remote sensing-based techniques are often insufficient to detect them, and they are still to be recognized and investigated in the field. Moreover, landslide deposits are more difficult to identify under the southern, forested monsoon Himalayan zone. In contrast, evidence for giant landslides is more prominent in areas with sparse or no vegetation such as that found above the tree-line like in Langtang (TsergoRi) and Kangchenjunga (Dzongri) areas (Weidinger and Korup 2009), or in the arid, rain shadow Transhimalayan area (Fort 2000; Mitchell et al. 2007; Dortch et al. 2009) and, along the two Himalayan syntaxes (Hewitt 2009; Korup et al. 2010), where hundreds of them were identified.

Location	Catchment	N	E	Landslid type	Rock type	Tectonic unit	Estimated volume (10^9 m^3)	Age	Ref.
Ringmo Suligad	Dolpo Nepal	29.18	82.94	RS	LI	TTH	1.5	$20,885 \pm 1675ka$ (^{36}Cl age)	(Fort et al. 2014)
Dhumpu-Chooya	Kali Gandaki Nepal	28.65	83.59	RS	GN	HHC	3	$29,674 \pm$ 1022 ka (^{10}Be age)	(Fort et al. 2014; Zech et al. 2009)
Braga	Marsyandi Nepal	28.38	84.4	RA / RS	MS	TTH	15	Pre-Last-Gla	(Fort 2011; Weidinger et al. 2014)
Pokhara	Seti khola Nepal	28.20	84.00	RA / DF	LI	TTH	3	pulses between 700 and 1700 AD (^{14}C age)	(Fort 1987b; Schwanghart et al. 2016; Stolle et al. 2017)
Latamrang	Marsyandi Nepal	28.53	84.31	RS	GN / QZ	HHC	5.5	5440 BP (^{14}C age)	(Korup et al. 2006; Pratt-Sitaula et al. 2007)
Tsergo Ri	Dudh Koshi, Nepal	28.25	85.58	RS	MS / GN / LG	HHC	10	25,000 to 40,000 BP (Fission track age)	(Weidinger et al. 2002; Takagi et al. 2007)

Table 4.2. *Selected characteristics of giant ($> 10^8$ m^3) catastrophic landslides in Nepal, modified and completed from Weidinger and Korup (2009). Landslide types: RA: rock avalanche; RS: rockslide; DF: debris flow. Rock type: MS: Metasedimentary; GN: Gneiss; LI: Limestone; QZ: Quartzite; LG: Leucogranite; Tectonic unit: HHC: Higher Himalayan Crystalline complex; TTH: Tibetan (Tethyan) metasediments*

Associated with slope steepness, high local relief and active tectonics, such giant landslides generally correspond to large, deep-seated rock-slope failures. They may result from the collapse of very high peaks (> 8,000 m in Langtang; (Fort et al. 1982); > 7,500 m in South Annapurna; (Fort 1987a; Schwanghart et al. 2016; Stolle et al. 2017), or from the collapse of a ridge (Dampu-Choya, Fort 2000; Weidinger et al. 2002). There is a strong influence of predesign features (Weidinger 2006), such as tectonic structures (main thrusts, dip-slopes sliding surfaces) or specific lithologies (gneisses, leucogranites; Korup and Weidinger 2011) favoring local relief > 2,000 m. Due to their large volumes, the giant landslide debris may run out over long distances, generally down valley as well as up valley and, when the valley is rather narrow, debris can run up the opposite slope, forming a "brandung" ridge. A general consequence is the damming of the valley and nearby rivers, hence impounding large lakes like in Ringmo (Fort et al. 2014, see Figure 4.3A), forcing upstream aggradation (Hewitt 2006; Korup et al. 2007) and, hiding the original valley floor with thick sheets of debris, as observed in Manang (Fort 2011) or Khumbu (Goetz et al. 2015). Sedimentary evidence of these giant deposits is generally characterized by their internal structure: a basal shattered matrix (frictionite) progressively passing up to the crushed, dislocated material where local sliding surfaces can be recognized, whereas the highest levels look like a carapace of huge blocks (Dortch et al. 2009; Weidinger and Korup 2009; Weidinger et al. 2014, see Figure 4.3B). These characteristics confirm an "en masse" displacement at high velocity (Schramm et al. 1998) and should avoid any confusion with either tectonic fault-zone breccia, or glacial till material, hence avoiding any misinterpretation on either tectonic or paleo-climatic potential causes of such material.

Though some form of preconditioning and long-term strength degradation of the failure plane is likely the main cause of these giant failures, the timing of their failure may eventually occur during events of (para)glacial debuttressing, pore-water pressure increase and/or seismic shaking, as evidenced by comparing deposit dating (see Tables 4.1 and 4.2) and paleo-glacial and paleo-seismic records. For example, the Ringmo rock-avalanche age (^{36}Cl) or Latamrang rockslide (^{14}C) dating may suggest a paraglacial collapse, though no such correlation can be established with the Dhampu–Choya rockslide (Fort et al. 2014) or the Keylong Serai rock-avalanche (Mitchell et al. 2007).

In contrast, the medieval ages (^{14}C) of the giant debris-flow pulses of Pokhara valley (Schwanghart et al. 2016; Stolle et al. 2017, 2019) match well the known records of medieval mega-earthquakes ($M_w > 8$, Bollinger et al. 2014). The smaller Hattian debris avalanche (80×10^6 m^3), triggered by the 2008 Kashmir earthquake, is actually a good example of such seismic shaking large failures (Owen et al. 2008).

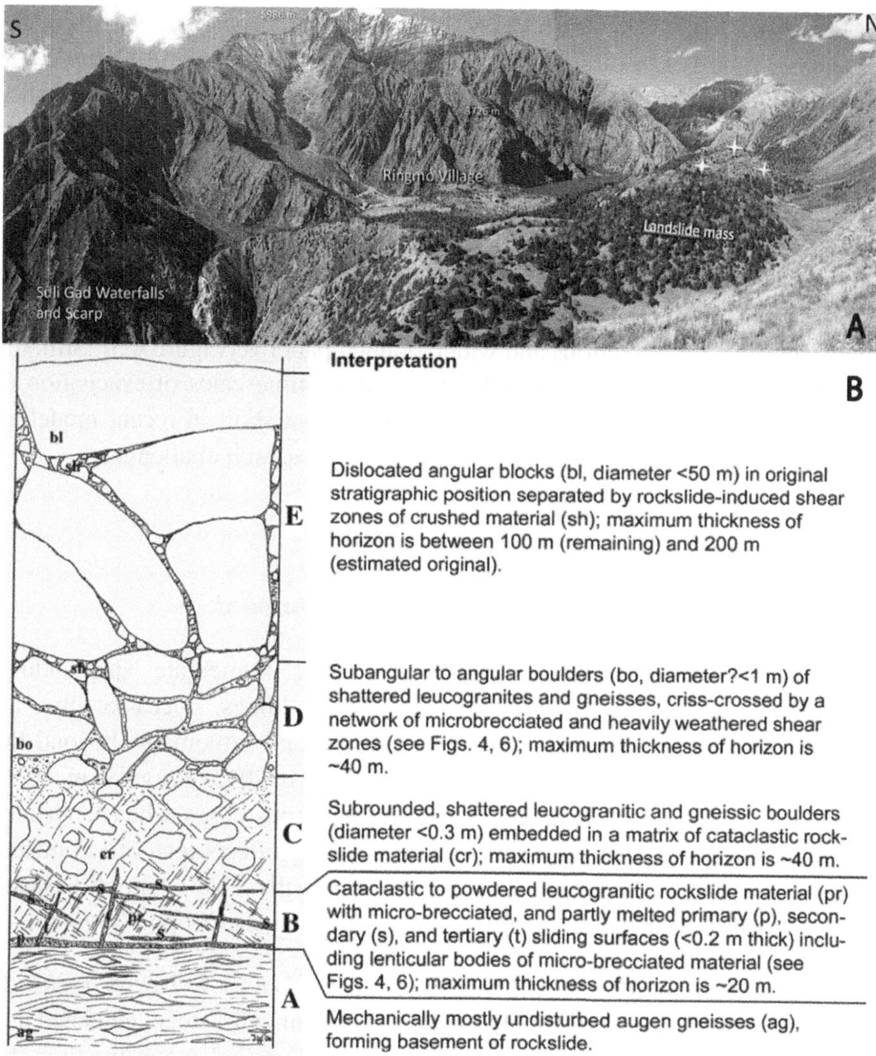

Figure 4.3. *View of the Ringmo giant landslide deposit (1.5 km³) which blocked the valley causing a lake upstream and a knickpoint downstream, Photo M. Fort 2010. B: typical sedimentary structure of giant landslide deposit, from (Weidinger and Korup 2009). For a color version of this figure, see www.iste.co.uk/cattin/himalaya3.zip*

Though infrequent, giant landslides significantly modulate the erosion of Himalayan topography. At a regional scale, they limit hillslope steepness and heights, and rapidly lower drainage divides (Korup and Weidinger 2011), as observed in the Annapurna Range at the Sabche La divide, and have been estimated to represent about 30% of the total long-term (10^5 yr) erosion budget (Marc et al. 2019). They also delay the sediment transport outward from the mountain zone (i.e. sediment storages and budgets; Bloethe and Korup 2013; Fort and Cossart 2013), creating a fragmented river system and a "disturbance regime" (Hewitt 2006), modifying the storage versus the sediment supply, independently of any tectonic or climatic variations. At a more local scale, they may delay river incision by burying the bedrock (Korup et al. 2010), potentially leading to channel abandonment and the formation of epigenetic gorges. They have also been argued to delay headward knickpoint erosion (Fort et al. 2010a; Korup et al. 2010), which explain the contrast in the river profiles between deep gorge sections and wider, braided patterns (Figure 4.3). Still, the controls on these giant landslide failures and the timescales of evacuation of their deposit, long according to field estimation but short in recent modeling studies (Croissant et al. 2017), remain important research challenges.

4.3. Landslides within Himalayan society

4.3.1. *Hazard cascades and their societal impact*

Mostly, landslides inflict damages in their immediate surrounding, destroying arable fields, infrastructure and buildings. Occasionally, air blasts may also significantly damage vegetation and settlements beyond the avalanche core runout, as was the case in Langtang in 2015 (Kargel et al. 2016; Gnyawali et al. 2020). However, there are numerous cases when landslides have triggered consecutive events with devastating effects tens of kilometers away from the landslide source. One such event is the 2021 Chamoli Disaster which began as landslide and turned into a cascading catastrophe (Figure 4.4).

The event took place on February 7, 2021, when a 27×10^6 m^3 rockslide detached from Ronti Peak, a 6,029 m high mountain with steep, glaciated and snow-covered flanks in the Garhwal Himalaya. Detailed reconstructions by Shugar et al. (2021) showed that the landslide mass consisted of rocks and glacier ice which melted by frictional heating as it descended downslope. The large amount of water transformed the rockslide into a fluid-rich and highly mobile debris flow traveling at maximum speeds of 25 m/s. Fifteen kilometer downstream of the source, the debris flow destroyed the Rishiganga hydropower project and then the Tapovan project another 10 km downriver (Figure 4.4). Of the 204 people killed or missing during the event, 190 were workers at the hydropower projects which faced a direct economic loss of \$223 million USD. Moreover, the effects of the event were felt hundreds of kilometers downstream as highly turbid water entered canals supplying Delhi with drinking water. High sediment concentrations are also expected to affect hydropower generation in the years to come as large amounts of sediments were intermediately deposited along the river and will be entrained during the upcoming floods.

In fact, hydropower is strongly affected by landslides. For example, Schwanghart et al. (2018) showed that the majority of damages to the Nepalese hydropower infrastructure during the Gorkha earthquake were inflicted indirectly by landslides rather than directly through ground shaking. A good predictor of where severe damages occur is river steepness (Wobus et al. 2006), a metric that highlights regions with high uplift rates and steep slopes (see Volume 1 – Chapter 6). Similarly, Devrani et al. (2015) showed that during Indian Floods 2013, the landslide-derived material was mostly reworked in steep river sections marking the transition to the High Himalaya, leading to either sediment burial or erosion of hydropower projects.

The socio-economic development of the Himalayan region largely depends on its infrastructure for power generation and transmission, traffic and communication. Landslides invoke high costs for the construction and maintenance of this infrastructure (Kumar et al. 2014), yet these costs often fall on impoverished communities (McAdoo et al. 2018). Reducing and managing the hazards and risks of landslides is thus pivotal to reduce the vulnerability of mountain communities, and to secure accessibility and energy availability.

4.3.2. *Human activities as an additional trigger of landslides*

Primary drivers of accelerated landslide activity in the Himalaya are anthropogenic interventions (Petley et al. 2007; Sidle and Ziegler 2012), including the rapid expansion of road networks, hydropower schemes, mines and quarries. Secondary drivers are rural irrigation, water supply and sanitation projects, migration trends and unmanaged hillslope developments (Leibundgut et al. 2016; Hearn and Shakya 2017).

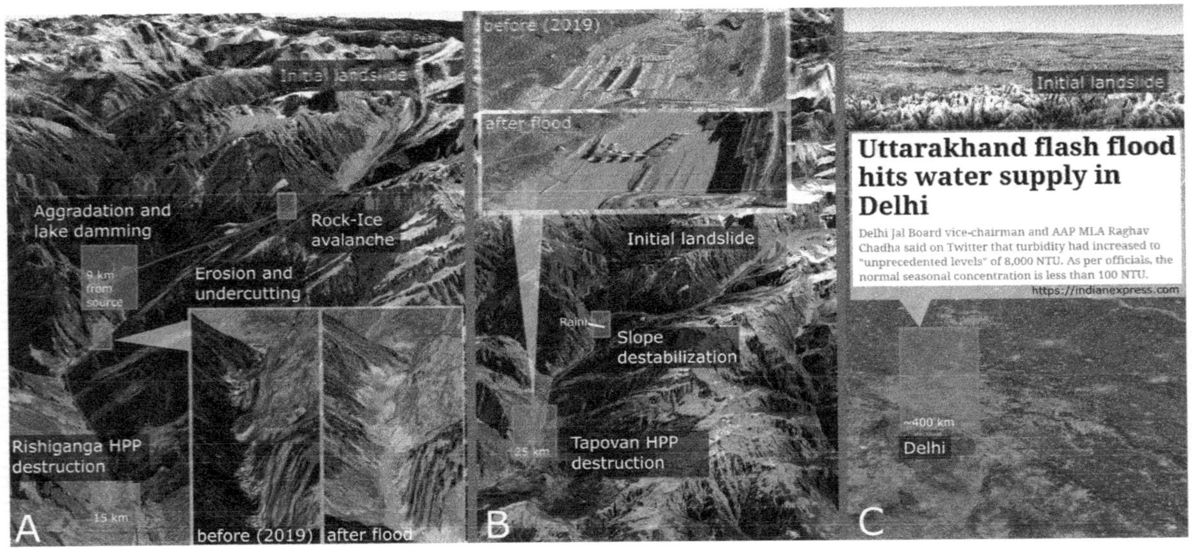

Figure 4.4. *The Chamoli Flood and its far-reaching effects. A) Within 15 km from its source, the landslide transformed to a debris avalanche and flow leading to profound geomorphological changes along its pathway. B) Lateral erosion led to slope destabilization in Raini and the flood wave to the destruction of a hydropower project 25 km downstream. C) Even 400 km away from the source, the landslide led to sustained increases in turbidity which affected drinking water supply in Delhi. Imagery from Google Earth. For a color version of this figure, see www.iste.co.uk/cattin/himalaya3.zip*

Road construction introduces four main disturbing agents on a natural slope: clearance of forest, excavation of hillsides, disposal of fill material and re-routing of natural drainage systems (Hearn and Shakya 2017, see Figure 4.5). In central Nepal, McAdoo et al. (2018) found that there are likely to be twice as many rainfall-triggered landslides in the terrain with poorly constructed roads as would be present without roads. In this region, hillside road construction involves a major cut of hills and partial filling. Only 15–30% of the total soil cut is used to fill, and the surplus is left on slopes, which is washed away by monsoon rainfall triggering debris flows and deposited into rivers (Talchabhadel et al. 2021).

The majority of landslides on roadsides of the Himalaya are shallow landslides and debris flow triggered by altering of the drainage channels, followed by re-activated monsoonal landslides. When road alignment crosses a tributary catchment partly or largely filled with colluvium (i.e. former landslide material stabilized by vegetation), the roadside cut destabilizes steep hillslopes, with progressive instability that causes the enlargement of past landslides, for example, landslide observed along the Kali Gandaki corridor (Bell et al. 2021). In addition, the opening of such roads in steeper and narrower gorge sections (in Higher Himalaya) requires rock excavations, causing rockslides and rockfalls, that may turn potentially into repeated rock collapses due to the absence of proper engineering mitigation and maintenance (Fort et al. 2010; Bell et al. 2021). In most cases, steep terrain is unavoidable in road alignments leading to less space for economical mitigations like ditches and deflection berms. Although guidelines for environment-friendly road construction techniques exist in countries like Nepal and India, they are not followed due to limited budget, pressure of access and lack of political interest towards the sustainability of these roads (Hearn and Shakya 2017; McAdoo et al. 2018). Increasing demand for energy is leading to the development of hydropower schemes (including large and small community-led projects), which are creating more landslides due to haphazard slope excavation and access road construction (Petley 2013). Growing population centers also indirectly enhance landslide activity because the nearby slopes are being excavated as rock quarries and mines that trigger large slope instabilities. Strict legal enforcement and policies are necessary to control these, but the prevailing poor governance in this region worsens the situation.

Migration trends linked with vanishing indigenous knowledge and cultural practices are also intricated with increased landslide activity in the Himalaya. Indeed, the abandonment of agricultural land on slopes traditionally stabilized by terracing efforts and control by village communities may lead to landslide reactivation, as observed in the South of the Annapurna range. Nearly 10% of abandoned irrigated rice fields were entirely damaged by landslides and floods (Khanal and Watanabe 2006). Furthermore, the hilly areas were vastly populated in the past and local communities used to clean stream channels and terraces before monsoons and major festivals, for example, "Kul Puja", "Navratri", etc. Nowadays, an increasing out-migration trend from those hillslopes into southward towns and cities exists (in search of better accessibility and opportunities). The once maintained water channels get clogged and shift the surficial and groundwater flow patterns, enhancing landslide activity. Such an increase in landslide activity may worsen with time in the absence of careful land management targeting vegetation (re)colonization and associated hillslope stabilization.

4.3.3. *Potential for mitigation*

Avoiding areas of high landslide susceptibility is the best way of reducing their impact, but this is often not possible in the Himalaya because of generalized steep topography, scattered mountain settlements and strong socio-economic-cultural ties with mountain landscapes. High frequency of landslides occurring during the monsoon season demands a quick response, field investigation and mitigation planning, which is often a challenge in this region.

High-tech and costly methods of landslide mitigation are rarely implemented because of the prevailing low- and middle-income countries in the region. Simple mitigation techniques involving community participation and low-to-medium tech engineered structures (e.g. masonry wall, surface drainage, gabion wall with horizontal drains, retaining walls, and rock anchorage, etc.) are implemented.

In the Himalayan regions, landslide mitigation can be categorized into three general measures (DSCWM 2016): (i) proper land husbandry through land-use improvement practices; (ii) managing water through proper drainage management and maintaining appropriate soil moisture, and (iii) strengthening slope stability through bio-engineering.

Slope bio-engineering, a nature-based solution, is a preferred choice in roadside slope stabilization (Figure 4.5). Bio-engineering incorporates the use of vegetation and engineering structures to increase slope stability. Vegetation increases the soil strength through its root structure, while the bio-engineered structure provides additional support. Bio-engineering does not replace standard civil engineering procedures, nor does it offer magic solutions. It does, however, increase the number of available ways to address roadside slope stability problems. Bio-engineering is especially useful in developing countries such as Nepal where it is relatively cheap to apply, given the locally available materials and the low cost of labor in this region (Devkota et al. 2014).

Road construction should proceed with proper drainage management and slope stabilization plans. In settlement areas, slope stability can be improved by proper management of surface water, where the community's local knowledge about the slope and drainage pattern is vital. Combining their knowledge of the slope with engineering design results in better drainage management plans to improve slope stability. In uninhabited lands, sediment control measures are vital to reduce the impact of cascading hazards like debris flows and floods.

However, the enforcement of existing legislation about safe techniques for road construction and drainage management for slope settlements remains a priority in order to avoid additional slope destabilization (McAdoo et al. 2018; Sudmeier-Rieux et al. 2019). More generally, before any major construction activity, slope hazard assessment should be strictly implemented, following land-use plans and policies.

Last but not least, in parallel with mitigation measures, national agencies can invest in early warning systems based on regional or local measurement systems (e.g. Thapa and Adhikari 2019; Cook et al. 2021), which, for a moderate cost, can provide alerts that could save hundreds of lives, as in the case of the hydropower plant workers who were killed in the Chamoli disaster (see above).

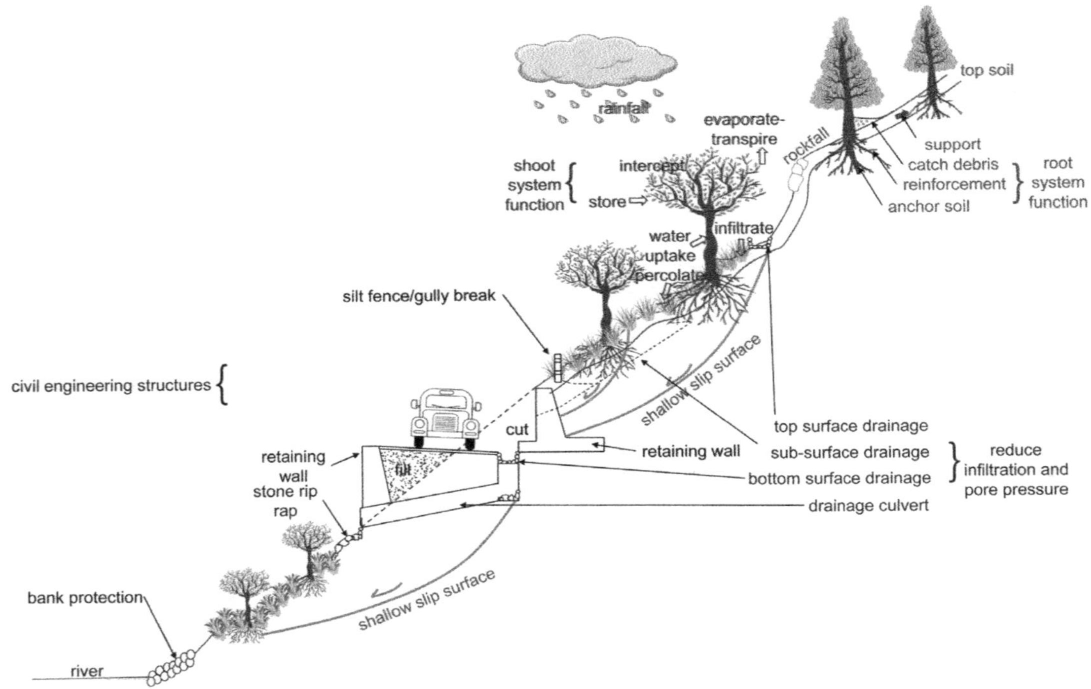

Figure 4.5. *Sketch showing the modification of slope during the construction of a road, and potential mitigation measures. Low-to-medium tech civil engineering structures in combination with engineered vegetation and proper drainage management plan can help enhance slope stability economically through bio-engineering*

4.3.4. *Climate change and future landsliding*

Climate change is expected to increase the mean global temperature by 1.4–5.8°C (see the 6th IPCC report), with an even greater impact on mountainous areas. Thus, the Himalaya's mean temperature may rise by about 5°C by the end of the 21st century (Sanjay et al. 2017), leading to changes in terms of rainfall regime, permafrost extent, flood occurrence and vegetation cover (Xu et al. 2007), which all affect slope stability and may increase landslide hazard (Gariano and Guzzetti 2016; Palmer 2020).

The primary driver for increasing landslide activity will likely be changes in rainfall patterns and monsoon timing. According to Sudmeier-Rieux et al. (2012), the Himalayan summer monsoon has become shorter, with more intense rainfall, and precipitation is expected to increase by roughly 20%. Though the nature of the changes remain debated, as others found the ongoing changes to be driven by more intense pre-monsoon rainfall (Brunello et al. 2020), most studies point towards rainfall intensification. Such intensification will enhance landslide frequency, as reported, for example, by Munoz-Torrero Manchado et al. (2021) in Western Nepal.

At larger and longer scales, coupling global circulation models and landslide empirical models suggest a widespread increase in the rate of landsliding, especially in places currently covered by glaciers and glacial lakes (Kirschbaum et al. 2020). Indeed, warmer climate will also impact snow melt dynamics and formation and drainage of glacial lakes, which, in combination with intense rainfall, may cause widespread landsliding, debris flows and bank collapses, with disastrous consequences, as in 2013 in Kedarnath (Allen et al. 2016), in 2018 in Kali Gandaki (Bell et al. 2021) or in 2021 in Melamchi. Rapid glaciers and permafrost melting under climate changes (Xu et al. 2007) are also increasing the number of glacial lakes (Veh et al. 2019; Furian et al. 2021). Thus, even if the frequency of GLOFs appears to have not increased yet (Veh et al. 2019), the ongoing growing abundance of glacial lakes may have important consequences for landslide hazard and its mitigation.

Finally, beyond changes affecting the meteorology and hydrology, climate change may also accelerate the rate of mass wasting by favoring ice and permafrost melting. This may substantially reduce the cohesion of steep, high elevation rock walls, leading to large landslides (Huggel et al. 2013) and may be at least in part responsible for the 2021 Chamoli disaster (Shugar et al. 2021). In conclusion, developing robust methods to predict the impact of climate change on landslide hazard and attribute specific events to ongoing climate change is an urgent research challenge, especially in the Himalaya.

4.4. Conclusion

In this chapter, we have reviewed the state of the art and current challenges faced by research on landslides in the Himalaya. A specificity of the Himalaya is to be exposed to frequent and high magnitude perturbations, such as strong shallow earthquakes, annual monsoon, localized extreme rainfall events and outburst floods. Each of these natural processes is an important cause of landsliding, but an emerging research challenge is also to understand how they are intertwined.

Future studies should explore questions such as: how are earthquakes influencing landsliding in the following monsoons? How are large landslides (whatever their triggers) able to cause outburst flood causing additional bank collapses? How to deconvolve the influence of high intensity rainfall events on hillslope mass wasting when they occur in the middle of sustained monsoonal rainfall? These questions will likely require a combination of cutting-edge approach in remote sensing and geophysical monitoring (from seismological to meteorological techniques), as well as innovative modeling approaches. Then, the question of what controls the occurrence and frequency of very large to giant landslides ($> 10^6 - 10^9$ m^3), numerous in the Himalayan range, is also a significant research challenge, with important implications for hazard and long-term erosion of the range. Additional field work, characterization and dating of large deposits and incorporation of this knowledge in geomechanical model are probably a way forward.

Beyond these questions on the natural drivers of landsliding in the Himalaya, we have also reviewed the substantial hazard that they constitute in the Himalaya. In particular, the questions of hazard cascade and perturbations causing prolonged conditions of elevated hazard, such as earthquakes, merit further investigation. Options for mitigation are also to be explored, keeping in mind that growing evidence suggests that some poorly planned human activities, such as road construction or land use change, contribute to slope instability and careful planning of such development activities should be seen as a key component of mitigation. We suggest that research involving social science is essential to understand how to make local communities aware of the risks and of solutions which they could appropriate. Last, given climate change is going to significantly change the conditions to which the Himalayan range is submitted, especially in terms of extreme precipitation and permafrost extent, special attention should be given to better constrain how landsliding depends on these factors and better model the evolution of these factors in the future decades.

4.5. References

Allen, S.K., Rastner, P., Arora, M., Huggel, C., Stoffel, M. (2016). Lake outburst and debris flow disaster at Kedarnath, June 2013: Hydrometeorological triggering and topographic predisposition. *Landslides*, 13(6), 1479–1491.

Ao, M., Zhang, L., Dong, Y., Su, L., Shi, X., Balz, T., Liao, M. (2020). Characterizing the evolution life cycle of the Sunkoshi landslide in Nepal with multi-source SAR data. *Scientific Reports*, 10(1), 17988.

Ashouri, H., Hsu, K.-L., Sorooshian, S., Braithwaite, D.K., Knapp, K.R., Cecil, L.D., Nelson, B.R., Prat, O.P. (2015). PERSIANN-CDR: Daily precipitation climate data record from multisatellite observations for hydrological and climate studies. *Bulletin of the American Meteorological Society*, 96(1), 69–83.

Avouac, J.-P. (2015). From geodetic imaging of seismic and aseismic fault slip to dynamic modeling of the seismic cycle. *Annual Review of Earth and Planetary Sciences*, 43(1), 233–271.

Bell, R., Fort, M., Goetz, J., Bernsteiner, H., Andermann, C., Etzlstorfer, J., Posch, E., Gurung, N., Gurung, S. (2021). Major geomorphic events and natural hazards during monsoonal precipitation 2018 in the Kali Gandaki Valley, Nepal Himalaya. *Geomorphology*, 372, 107451.

Bloethe, J.H. and Korup, O. (2013). Millennial lag times in the Himalayan sediment routing system. *Earth and Planetary Science Letters*, 382, 38–46.

Bogaard, T.A. and Greco, R. (2016). Landslide hydrology: From hydrology to pore pressure. *WIREs Water*, 3(3), 439–459.

Bollinger, L., Sapkota, S.N., Tapponnier, P., Klinger, Y., Rizza, M., Van der Woerd, J., Tiwari, D.R., Pandey, R., Bitri, A., Bes de Berc, S. (2014). Estimating the return times of great Himalayan earthquakes in eastern Nepal: Evidence from the Patu and Bardibas strands of the Main Frontal Thrust. *Journal of Geophysical Research: Solid Earth*, 119(9), 2014JB010970.

Bookhagen, B., Thiede, R.C., Strecker, M.R. (2005). Late Quaternary intensified monsoon phases control landscape evolution in the northwest Himalaya. *Geology*, 33(2), 149–152.

Burrows, K., Marc, O., Remy, D. (2022). Using Sentinel-1 radar amplitude time series to constrain the timings of individual landslides: A step towards understanding the controls on monsoon-triggered landsliding. *Natural Hazards and Earth System Sciences*, 22(8), 2637–2653. https://doi.org/10.5194/nhess-22-2637-2022.

Brunello, C.F., Andermann, C., Marc, O., Schneider, K.A., Comiti, F., Achleitner, S., Hovius, N. (2020). Annually resolved monsoon onset and withdrawal dates across the Himalayas derived from local precipitation statistics. *Geophysical Research Letters*, 47(23), e2020GL088420.

Byers, A.C., Rounce, D.R., Shugar, D.H., Lala, J.M., Byers, E.A., Regmi, D. (2019). A rockfall-induced glacial lake outburst flood, Upper Barun Valley, Nepal. *Landslides*, 16(3), 533–549.

Chakraborty, I., Ghosh, D.S., Bhattacharya, D., Bora, A. (2014). Earthquake induced landslides in the Sikkim-Darjeeling Himalayas – An aftermath of the 18th September 2011 Sikkim earthquake. Report, Geological Survey of India.

Chigira, M., Tsou, C.-Y., Matsushi, Y., Hiraishi, N., Matsuzawa, M. (2013). Topographic precursors and geological structures of deep-seated catastrophic landslides caused by Typhoon Talas. *Geomorphology*, 201, 479–493. https://doi.org/10.1016/j.geomorph.2013.07.020.

Cook, K.L., Andermann, C., Gimbert, F., Adhikari, B.R., Hovius, N. (2018). Glacial lake outburst floods as drivers of fluvial erosion in the Himalaya. *Science*, 362(6410), 53–57.

Cook, K.L., Rekapalli, R., Dietze, M., Pilz, M., Cesca, S., Rao, N.P., Srinagesh, D., Paul, H., Metz, M., Mandal, P. et al. (2021). Detection and potential early warning of catastrophic flow events with regional seismic networks. *Science*, 374(6563), 87–92.

Croissant, T., Lague, D., Steer, P., Davy, P. (2017). Rapid post-seismic landslide evacuation boosted by dynamic river width. *Nature Geoscience*, 10(9), 680–684.

Dahal, R.K. and Hasegawa, S. (2008). Representative rainfall thresholds for landslides in the Nepal Himalaya. *Geomorphology*, 100(3–4), 429–443.

Dahlquist, M.P. and West, A.J. (2019). Initiation and runout of post-seismic debris flows: Insights from the 2015 Gorkha earthquake. *Geophysical Research Letters*, 46(16), 9658–9668.

De Blasio, F.V. (2011). *Introduction to the Physics of Landslides: Lecture Notes on the Dynamics of Mass Wasting*. Springer, Dordrecht.

DeGraff, J.V., Cannon, S.H., Gartner, J.E. (2015). The timing of susceptibility to post-fire debris flows in the Western United States. *Environmental and Engineering Geoscience*, 21(4), 277–292.

Devkota, S., Sudmeier-Rieux, K., Penna, I., Eberle, S., Jaboyedoff, M., Adhikari, A., Khanal, R. (2014). *Community Based Bioengineering for Ecosafe Roadsides in Nepal.* University of Lausanne, International Union for Conservation of Nature, Nepal and Department of Soil Conservation and Watershed Management, Government of Nepal.

Devrani, R., Singh, V., Mudd, S.M., Sinclair, H.D. (2015). Prediction of flash flood hazard impact from Himalayan river profiles. *Geophysical Research Letters*, 42(14), 5888–5894.

Dikshit, A., Sarkar, R., Pradhan, B., Acharya, S., Dorji, K. (2019). Estimating rainfall thresholds for landslide occurrence in the Bhutan Himalayas. *Water*, 11(8), 1616.

Dimri, A.P., Chevuturi, A., Niyogi, D., Thayyen, R.J., Ray, K., Tripathi, S.N., Pandey, A.K., Mohanty, U.C. (2017). Cloudbursts in Indian Himalayas: A review. *Earth-Science Reviews*, 168, 1–23.

Dortch, J.M., Owen, L.A., Haneberg, W.C., Caffee, M.W., Dietsch, C., Kamp, U. (2009). Nature and timing of large landslides in the Himalaya and Transhimalaya of Northern India. *Quaternary Science Reviews*, 28(11–12), 1037–1054.

DSCWM (2016). *Guideline on Landslide Treatment and Mitigation.* DSCWM, Kathmandu.

Dunham, A.M., Kiser, E., Kargel, J.S., Haritashya, U.K., Watson, C.S., Shugar, D.H., Hughes, A., DeCelles, P.G. (2022). Topographic control on ground motions and landslides from the 2015 Gorkha earthquake. *Geophysical Research Letters*, 49(10), e2022GL098582. https://doi.org/10.1029/2022GL098582.

Fort, M. (1987a). Geomorphic and hazards mapping in the dry, continental Himalaya: 1:50,000 maps of Mustang District, Nepal. *Mountain Research and Development*, 7, 222–238.

Fort, M. (1987b). Sporadic morphogenesis in a continental subduction setting. *Zeitschrift für Geomorphologie N.F.*, 63(Suppl.), 9–36.

Fort, M. (2000). Glaciers and mass wasting processes: Their influence on the shaping of the Kali Gandaki Valley (Higher Himalaya of Nepal). *Quaternary International*, 65(3), 101–119.

Fort, M. (2011). The Himalayas: From mountain building to landform evolution in a changing world. *Geographica Polonica*, 84, 15–37.

Fort, M. and Cossart, E. (2013). Erosion assessment in the Middle Kali Gandaki (Nepal): A sediment budget approach. *Journal of Nepal Geological Society*, 46, 25–40.

Fort, M., Freytet, P., Colchen, M. (1982). Structural and sedimentological evolution of the Thakkhola Mustang Graben (Nepal Himalayas). *Zeitschrift für Geomorphologie, N.F.*, 42(Suppl.), 75–98.

Fort, M., Cossart, E., Arnaud-Fassetta, G. (2010a). Catastrophic landslides and sedimentary budgets. In *Geomorphological Hazards and Disaster Prevention*, Alcantara-Ayala, I. and Goudie, A. (eds). Cambridge University Press, Cambridge.

Fort, M., Cossart, E., Arnaud-Fassetta, G. (2010b). Hillslope-channel coupling in the Nepal Himalayas and threat to man-made structures: The Middle Kali Gandaki Valley. *Geomorphology*, 124(3–4), 178–199.

Fort, M., Braucher, R., Bourles, D., Guillou, V., Nath Rimal, L., Gribenski, N., Cossart, E. (2014). Geomorphic impacts, age and significance of two giant landslide dams in the Nepal Himalayas: Ringmo-Phoksundo (Dolpo District) and Dhampu-Chhoya (Mustang District). *EGU General Assembly Conference Abstracts*, 16, 5615.

Furian, W., Loibl, D., Schneider, C. (2021). Future glacial lakes in High Mountain Asia: An inventory and assessment of hazard potential from surrounding slopes. *Journal of Glaciology*, 67(264), 653–670.

Gabet, E.J., Burbank, D.W., Putkonen, J.K., Pratt-Sitaula, B.A., Ojha, T. (2004). Rainfall thresholds for landsliding in the Himalayas of Nepal. *Geomorphology*, 63(3–4), 131–143.

Gallen, S.F., Clark, M.K., Godt, J.W., Roback, K., Niemi, N.A. (2017). Application and evaluation of a rapid response earthquake-triggered landslide model to the 25 April 2015 Mw 7.8 Gorkha Earthquake, Nepal. *Tectonophysics*, 714–715(Supplement C), 173–187.

Gariano, S.L. and Guzzetti, F. (2016). Landslides in a changing climate. *Earth-Science Reviews*, 162, 227–252.

Gnyawali, K.R. and Adhikari, B.R. (2017). Spatial relations of earthquake induced landslides triggered by 2015 Gorkha Earthquake Mw = 7.8. In *Advancing Culture of Living with Landslides*, Mikoš, M., Casagli, N., Yin, Y. Sassa, K. (eds). Springer International Publishing, Cham.

Gnyawali, K.R., Xing, A., Zhuang, Y. (2020). Dynamic analysis of the multi-staged ice-rock debris avalanche in the Langtang valley triggered by the 2015 Gorkha earthquake, Nepal. *Engineering Geology*, 265, 105440.

Goetz, J., Weidinger, J.T., Kraxberger, S., Hennecke, A.-L., Buckel, J., Adhikari, B.R. (2015). Geomorphologic and hydrogeologic characteristics of populated rockslide deposits (Sagarmatha National Park, Khumbu Himal, Nepal). *Journal of Water Resource and Protection*, 7(13), 1038.

Handwerger, A.L., Rempel, A.W., Skarbek, R.M., Roering, J.J., Hilley, G.E. (2016). Rate-weakening friction characterizes both slow sliding and catastrophic failure of landslides. *Proceedings of the National Academy of Sciences*, 113(37), 10281–10286.

Hearn, G.J. and Shakya, N.M. (2017). Engineering challenges for sustainable road access in the Himalayas. *Quarterly Journal of Engineering Geology and Hydrogeology*, 50(1), 69–80.

Hewitt, K. (2006). Disturbance regime landscapes: Mountain drainage systems interrupted by large rockslides. *Progress in Physical Geography*, 30, 365–393.

Hewitt, K. (2009). Rock avalanches that travel onto glaciers and related developments, Karakoram Himalaya, Inner Asia. *Geomorphology*, 103(1), 66–79.

Hobley, D.E.J., Sinclair, H.D., Mudd, S.M. (2012). Reconstruction of a major storm event from its geomorphic signature: The Ladakh floods, 6 August 2010. *Geology*, 40(6), 483–486.

Huggel, C., Gruber, S., Korup, O. (2013). 13.17 Landslide hazards and climate change in high mountains. In *Treatise on Geomorphology*, Shroder, J.F. (ed.). Academic Press, San Diego.

Iverson, R.M. (2000). Landslide triggering by rain infiltration. *Water Resources Research*, 36(7), 1897–1910.

Jones, J.N., Boulton, S.J., Stokes, M., Bennett, G.L., Whitworth, M.R.Z. (2021). 30-year record of Himalaya mass-wasting reveals landscape perturbations by extreme events. *Nature Communications*, 12(1), 6701.

Kargel, J.S., Leonard, G.J., Shugar, D.H., Haritashya, U.K., Bevington, A., Fielding, E.J., Fujita, K., Geertsema, M., Miles, E.S., Steiner, J. et al. (2016). Geomorphic and geologic controls of geohazards induced by Nepal's 2015 Gorkha earthquake. *Science*, 351(6269), aac8353 [Online]. Available at: https://www.science.org/doi/10.1126/science.aac8353.

Keefer, D.K. (1984). Landslides caused by earthquakes. *Geological Society of America Bulletin*, 95(4), 406–421.

Keefer, D.K. (1994). The importance of earthquake-induced landslides to long-term slope erosion and slope-failure hazards in seismically active regions. *Geomorphology*, 10(1–4), 265–284.

Khanal, N.R. and Watanabe, T. (2006). Abandonment of agricultural land and its consequences. *Mountain Research and Development*, 26(1), 32–40.

Khattak, G.A., Owen, L.A., Kamp, U., Harp, E.L. (2010). Evolution of earthquake-triggered landslides in the Kashmir Himalaya, northern Pakistan. *Geomorphology*, 115(1–2), 102–108.

Kirschbaum, D., Kapnick, S.B., Stanley, T., Pascale, S. (2020). Changes in extreme precipitation and landslides over High Mountain Asia. *Geophysical Research Letters*, 47(4), e2019GL085347.

Korup, O. and Weidinger, J.T. (2011). Rock type, precipitation, and the steepness of Himalayan threshold hillslopes. *Geological Society, London, Special Publications*, 353(1), 235–249.

Korup, O., Strom, A., Weidinger, J.T. (2006). Fluvial response to large rock-slope failures: Examples from the Himalayas, the Tien Shan, and the Southern Alps in New Zealand. *Geomorphology*, 78, 3–21.

Korup, O., Clague, J., Hermanns, R., Hewitt, K., Strom, A., Weidinger, J.T. (2007). Giant landslides, topography, and erosion. *Earth and Planetary Science Letters*, 26, 578–589.

Korup, O., Montgomery, D.R., Hewitt, K., Burchfiel, B.C. (2010). Glacier and landslide feedbacks to topographic relief in the Himalayan syntaxes. *Proceedings of the National Academy of Sciences of the United States of America*, 107(12), 5317–5322.

Kumar, K., Jangpangi, L., Gangopadhyay, S. (2014). Highways vs. landslides and their consequences in Himalaya. In *Landslide Science for a Safer Geoenvironment*, Sassa, K., Canuti, P., Yin, Y. (eds). Springer International Publishing, Cham.

Kumar, A., Asthana, A., Priyanka, R.S., Jayangondaperumal, R., Gupta, A.K., Bhakuni, S. (2017). Assessment of landslide hazards induced by extreme rainfall event in Jammu and Kashmir Himalaya, northwest India. *Geomorphology*, 284, 72–87.

Lacroix, P. and Amitrano, D.R. (2013). Long-term dynamics of rockslides and damage propagation inferred from mechanical modeling. *Journal of Geophysical Research: Earth Surface*, 118(4), 2292–2307.

Larsen, I.J. and Montgomery, D.R. (2012). Landslide erosion coupled to tectonics and river incision. *Nature Geoscience*, 5(7), 468–473.

Larsen, I., Montgomery, D.R., Korup, O. (2010). Landslide erosion controlled by hillslope material. *Nature Geoscience*, 3(4), 247–251.

Leibundgut, G., Sudmeier-Rieux, K., Devkota, S., Jaboyedoff, M., Derron, M.-H., Penna, I., Nguyen, L. (2016). Rural earthen roads impact assessment in Phewa watershed, Western region, Nepal. *Geoenvironmental Disasters*, 3(1), 13.

Marc, O., Hovius, N., Meunier, P., Uchida, T., Hayashi, S. (2015). Transient changes of landslide rates after earthquakes. *Geology*, 43, 883–886.

Marc, O., Hovius, N., Meunier, P., Gorum, T., Uchida, T. (2016). A seismologically consistent expression for the total area and volume of earthquake-triggered landsliding. *Journal of Geophysical Research: Earth Surface*, 121(4), 640–663.

Marc, O., Meunier, P., Hovius, N. (2017). Prediction of the area affected by earthquake-induced landsliding based on seismological parameters. *Natural Hazards and Earth System Sciences*, 17(7), 1159–1175.

Marc, O., Behling, R., Andermann, C., Turowski, J.M., Illien, L., Roessner, S., Hovius, N. (2019). Long-term erosion of the Nepal Himalayas by bedrock landsliding: The role of monsoons, earthquakes and giant landslides. *Earth Surface Dynamics*, 7(1), 107–128.

Marc, O., Sens-Schoenfelder, C., Illien, L., Meunier, P., Hobiger, M., Sawazaki, K., Rault, C., Hovius, N. (2021). Toward using seismic interferometry to quantify landscape mechanical variations after earthquakes. *Bulletin of the Seismological Society of America*, 111(3), 1631–1649.

Martha, T.R., Roy, P., Mazumdar, R., Govindharaj, K.B., Kumar, K.V. (2017). Spatial characteristics of landslides triggered by the 2015 Mw 7.8 (Gorkha) and Mw 7.3 (Dolakha) earthquakes in Nepal. *Landslides*, 14, 697–704. https://doi.org/10.1007/s10346-016-0763-x.

Mathew, J., Babu, D.G., Kundu, S., Kumar, K.V., Pant, C.C. (2014). Integrating intensity-duration-based rainfall threshold and antecedent rainfall-based probability estimate towards generating early warning for rainfall-induced landslides in parts of the Garhwal Himalaya, India. *Landslides*, 11(4), 575–588.

Mathur, L. (1953). Assam earthquake of 15th August 1950, a short note on factual observations. *A Compilation of Papers on the Assam Earthquake of August, 15, 1950*, M.B Ramachandra Rao edition, Vol. 1. The Central Board of Geophysical Publisher, National Geophysical Research Institute, Hyderabad.

McAdoo, B.G., Quak, M., Gnyawali, K.R., Adhikari, B.R., Devkota, S., Rajbhandari, P.L., Sudmeier-Rieux, K. (2018). Roads and landslides in Nepal: How development affects environmental risk. *Natural Hazards and Earth System Sciences*, 18(12), 3203–3210.

Meunier, P., Hovius, N., Haines, A.J. (2007). Regional patterns of earthquake-triggered landslides and their relation to ground motion. *Geophysical Research Letters*, 34(20), L20408.

Meunier, P., Uchida, T., Hovius, N. (2013). Landslide patterns reveal the sources of large earthquakes. *Earth and Planetary Science Letters*, 363, 27–33.

Milledge, D.G., Bellugi, D., McKean, J.A., Densmore, A.L., Dietrich, W.E. (2014). A multidimensional stability model for predicting shallow landslide size and shape across landscapes. *Journal of Geophysical Research: Earth Surface*, 119(11), 2014JF003135.

Mitchell, W.A., McSaveney, M.J., Zondervan, A., Kim, K., Dunning, S.A., Taylor, P.J. (2007). The Keylong Serai rock avalanche, NW Indian Himalaya: Geomorphology and palaeoseismic implications. *Landslides*, 4(3), 245–254.

Munoz-Torrero Manchado, A., Allen, S., Ballesteros-Cánovas, J.A., Dhakal, A., Dhital, M.R., Stoffel, M. (2021). Three decades of landslide activity in western Nepal: New insights into trends and climate drivers. *Landslides*, 18(6), 2001–2015.

Newmark, N. (1965). Effects of earthquakes on dams and embankments. *Geotechnique*, 15(2), 139–159.

Owen, L.A., Kamp, U., Khattak, G.A., Harp, E.L., Keefer, D.K., Bauer, M.A. (2008). Landslides triggered by the 8 October 2005 Kashmir earthquake. *Geomorphology*, 94(1–2), 1–9.

Palmer, J. (2020). *A Slippery Slope: Could Climate Change Lead to More Landslides?* Eos, Washington, D.C.

Parajuli, A., Gautam, A.P., Sharma, S.P., Bhujel, K.B., Sharma, G., Thapa, P.B., Bist, B.S., Poudel, S. (2020). Forest fire risk mapping using GIS and remote sensing in two major landscapes of Nepal. *Geomatics, Natural Hazards and Risk*, 11(1), 2569–2586.

Petley, D.N. (2013). Global losses from landslides associated with dams and reservoirs. *Italian Journal of Engineering Geology and Environment*, 6, 63–72.

Petley, D.N., Higuchi, T., Petley, D.J., Bulmer, M.H., Carey, J. (2005). Development of progressive landslide failure in cohesive materials. *Geology*, 33(3), 201–204.

Petley, D.N., Hearn, G.J., Hart, A., Rosser, N.J., Dunning, S.A., Oven, K., Mitchell, W.A. (2007). Trends in landslide occurrence in Nepal. *Natural Hazards*, 43(1), 23–44.

Pratt-Sitaula, B., Garde, M., Burbank, D.W., Oskin, M., Heimsath, A., Gabet, E. (2007). Bedload-to-suspended load ratio and rapid bedrock incision from Himalayan Landslide-dam lake record. *Quaternary Research*, 68(1), 111–120.

Roback, K., Clark, M.K., West, A.J., Zekkos, D., Li, G., Gallen, S.F., Chamlagain, D., Godt, J.W. (2018). The size, distribution, and mobility of landslides caused by the 2015 Mw7.8 Gorkha earthquake, Nepal. *Geomorphology*, 301, 121–138.

Rosser, N., Kincey, M., Oven, K., Densmore, A., Robinson, T., Pujara, D.S., Shrestha, R., Smutny, J., Gurung, K., Lama, S. et al. (2021). Changing significance of landslide hazard and risk after the 2015 Mw 7.8 Gorkha, Nepal Earthquake. *Progress in Disaster Science*, 10, 100159.

Saba, S.B., van der Meijde, M., van der Werff, H. (2010). Spatiotemporal landslide detection for the 2005 Kashmir earthquake region. *Geomorphology*, 124(1–2), 17–25.

Sanjay, J., Krishnan, R., Shrestha, A.B., Rajbhandari, R., Ren, G.-Y. (2017). Downscaled climate change projections for the Hindu Kush Himalayan region using CORDEX South Asia regional climate models. *Advances in Climate Change Research*, 8(3), 185–198.

Schramm, J.M., Weidinger, J.T., Ibetsberger, H.J. (1998). Petrologic and structural controls on geomorphology of prehistoric Tsergo Ri slope failure, Langtang Himal, Nepal. *Geomorphology*, 26(1), 107–121.

Schwanghart, W., Bernhardt, A., Stolle, A., Hoelzmann, P., Adhikari, B.R., Andermann, C., Tofelde, S., Merchel, S., Rugel, G., Fort, M. et al. (2016). Repeated catastrophic valley infill following medieval earthquakes in the Nepal Himalaya. *Science*, 351(6269), 147–150.

Schwanghart, W., Ryan, M., Korup, O. (2018). Topographic and seismic constraints on the vulnerability of Himalayan hydropower. *Geophysical Research Letters*, 45(17), 8985–8992.

Shroder Jr., J.F. (1998). Slope failure and denudation in the western Himalaya. *Geomorphology*, 26(1–3), 81–105.

Shugar, D.H., Jacquemart, M., Shean, D., Bhushan, S., Upadhyay, K., Sattar, A., Schwanghart, W., McBride, S., de Vries, M.V.W., Mergili, M. et al. (2021). A massive rock and ice avalanche caused the 2021 disaster at Chamoli, Indian Himalaya. *Science*, 373(6552), 300–306.

Sidle, R.C. and Ziegler, A.D. (2012). The dilemma of mountain roads. *Nature Geoscience*, 5(7), 437–438.

Stevens, V.L. and Avouac, J.-P. (2016). Millenary Mw > 9.0 earthquakes required by geodetic strain in the Himalaya. *Geophysical Research Letters*, 43(3), 2015GL067336.

Stolle, A., Bernhardt, A., Schwanghart, W., Hoelzmann, P., Adhikari, B.R., Fort, M., Korup, O. (2017). Catastrophic valley fills record large himalayan earthquakes, Pokhara, Nepal. *Quaternary Science Reviews*, 177(Supplement C), 88–103.

Stolle A., Schwanghart W., Andermann C., Bernhardt A., Fort M., Jansen J., Wittmann H., Merchel S., Rugel G., Adhikari B.R., Korup O. (2019). Protracted river response to medieval earthquakes. *Earth Surface Processes and Landforms*, 44, 331–341.

Sudmeier-Rieux, K., Gaillard, J.-C., Sharma, S., Dubois, J., Jaboyedoff, M. (2012). Chapter 7: Floods, landslides, and adapting to climate change in Nepal: What role for climate change models? In *Climate Change Modeling for Local Adaptation in the Hindu Kush-Himalayan Region*, Lamadrid, A. and Kelman, I. (eds). Emerald Group Publishing Limited, Bingley.

Sudmeier-Rieux, K., McAdoo, B.G., Devkota, S., Rajbhandari, P.C.L., Howell, J., Sharma, S. (2019). Invited perspectives: Mountain roads in Nepal at a new crossroads. *Natural Hazards and Earth System Sciences*, 19(3), 655–660.

Takagi, H., Arita, K., Danhara, T., Iwano, H. (2007). Timing of the Tsergo Ri landslide, Langtang Himal, determined by fission-track dating of pseudotachylyte. *Journal of Asian Earth Sciences*, 29(2), 466–472.

Talchabhadel, R., Panthi, J., Sharma, S., Ghimire, G.R., Baniya, R., Dahal, P., Baniya, M.B., K.C., S., Jha, B., Kaini, S. et al. (2021). Insights on the impacts of hydroclimatic extremes and anthropogenic activities on sediment yield of a river basin. *Earth*, 2(1), 32–50.

Thapa, P.S. and Adhikari, B.R. (2019). Development of community-based landslide early warning system in the earthquake-affected areas of Nepal Himalaya. *Journal of Mountain Science*, 16(12), 2701–2713.

Tian, Y., Owen, L., Xu, C., Shen, L., Zhou, Q., Figueiredo, P. (2020). Geomorphometry and statistical analyses of landslides triggered by the 2015 Mw 7.8 Gorkha earthquake and the Mw 7.3 aftershock, Nepal. *Frontiers in Earth Science*, 8, 407.

Valagussa, A., Frattini, P., Valbuzzi, E., Crosta, G.B. (2021). Role of landslides on the volume balance of the Nepal 2015 earthquake sequence. *Scientific Reports*, 11(1), 3434.

Van Asch, T.W.J., Buma, J., Van Beek, L.P.H. (1999). A view on some hydrological triggering systems in landslides. *Geomorphology*, 30(1), 25–32.

Veh, G., Korup, O., von Specht, S., Roessner, S., Walz, A. (2019). Unchanged frequency of moraine dammed glacial lake outburst floods in the Himalaya. *Nature Climate Change*, 9, 379–383.

Weidinger, J.T. (2006). Predesign, failure and displacement mechanisms of large rockslides in the Annapurna Himalayas, Nepal. *Engineering Geology*, 83, 201–216.

Weidinger, J.T. and Korup, O. (2009). Frictionite as evidence for a large Late Quaternary rockslide near Kanchenjunga, Sikkim Himalayas, India – Implications for extreme events in mountain relief destruction. *Geomorphology*, 103(1), 57–65.

Weidinger, J.T., Schramm, J.-M., Nuschej, F. (2002). Ore mineralization causing slope failure in a high altitude mountain crest – On the collapse of an 8000m peak in Nepal. *Journal of Asian Earth Sciences*, 21(3), 295–306.

Weidinger, J.T., Korup, O., Munack, H., Altenberger, U., Dunning, S.A., Tippelt, G., Lottermoser, W. (2014). Giant rockslides from the inside. *Earth and Planetary Science Letters*, 389(Supplement C), 62–73.

Wobus, C., Whipple, K.X., Kirby, E., Snyder, N., Johnson, J., Spyropolou, K., Crosby, B., Sheehan, D. (2006). Tectonics from topography: Procedures, promise, and pitfalls. *GSA Special Papers*, 398, 55–74.

Xu, J., Eriksson, M., Vaidya, R., Shrestha, A.B., Hewitt, K. (2007). *The Melting Himalayas: Regional Challenges and Local Impacts of Climate Change on Mountain Ecosystems and Livelihoods.* International Centre for Integrated Mountain Development (ICIMOD) edition, Kathmandu.

Yatagai, A., Kamiguchi, K., Arakawa, O., Hamada, A., Yasutomi, N., Kitoh, A. (2012). APHRODITE: Constructing a long-term daily gridded precipitation dataset for Asia based on a dense network of rain gauges. *Bulletin of the American Meteorological Society*, 93(9), 1401–1415.

Yuan, R.-M., Deng, Q.-H., Cunningham, D., Xu, C., Xu, X.-W., Chang, C.-P. (2013). Density distribution of landslides triggered by the 2008 Wenchuan earthquake and their relationships to peak ground acceleration. *Bulletin of the Seismological Society of America*, 103(4), 2344–2355.

Zech, R., Zech, M., Kubik, P.W., Kharki, K., Zech, W. (2009). Deglaciation and landscape history around Annapurna, Nepal, based on Be surface exposure dating. *Quaternary Science Reviews*, 28(11–12), 1106–1118.

5

Himalayan Surface Rupturing Earthquakes

Laurent BOLLINGER[1], Matthieu FERRY[2], Romain LE ROUX-MALLOUF[3], Jérôme VAN DER WOERD[4] and Yann KLINGER[5]

[1] *French Alternative Energies and Atomic Energy Commission, Bruyères-le-Châtel, France*

[2] *University of Montpellier, France*

[3] *EDF, Aix-en-Provence, France*

[4] *University of Strasbourg, France*

[5] *Paris Cité University, France*

5.1. Introduction

The Himalayan range is affected by very strong, devastating earthquakes that have been mentioned in historical chronicles on numerous occasions since the medieval period. Recently, surface ruptures from several of these large historical earthquakes have been discovered in the field. Their surveys allowed us to estimate the spatial extent of the ruptured fault segments as well as the amount of slip caused by the earthquakes. Seismic surface ruptures along the Himalayan frontal thrust are also accompanied by abandonment of alluvial terraces. The joint study of these terraces and of paleoseismological

Himalaya, Dynamics of a Giant 3,
coordinated by Rodolphe CATTIN and Jean-Luc EPARD.

excavations through the ruptured faults provided a 5,000-year long history of the largest events, leading to the determination of their return time.

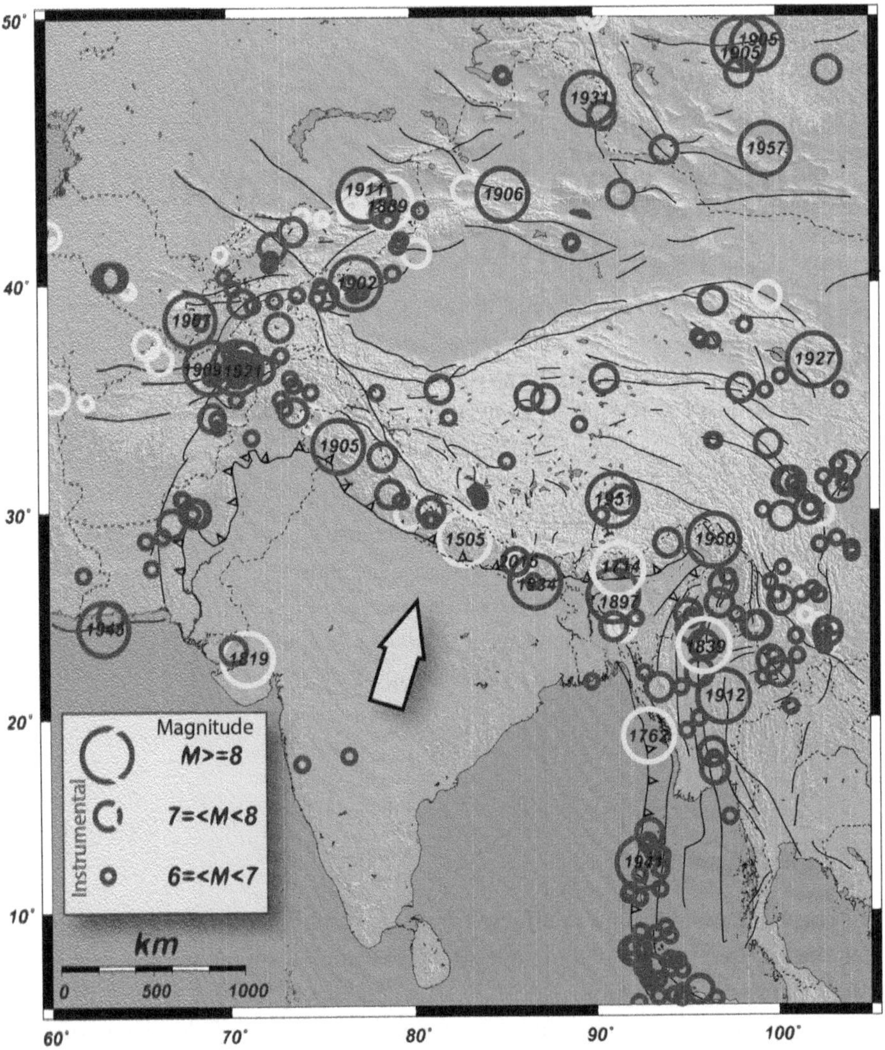

Figure 5.1. *Epicenters of large, damaging earthquakes compiled from the ISC-GEM catalogue and complementary sources. Historical events are shown in yellow; instrumental events are shown in red. For the largest events, the year is shown. The gray arrow indicates the direction of the India– Asia convergence. For a color version of this figure, see www.iste.co.uk/cattin/himalaya3.zip*

5.2. The large devastating earthquakes in the Himalaya

Strong devastating earthquakes affect the entire Himalayan range (Figure 5.1). These earthquakes result from plate tectonics (see Volume 1 – Chapter 1), and more specifically from the India–Asia convergence. This total convergence is close to 4 cm/year, and nearly half of it is accommodated through the Himalaya. The Himalayan shortening of the order of 2 cm/year appears to be essentially accommodated by the Main Himalayan Thrust, the great thrust fault that separates the India plate which plunges below the Himalaya and Tibetan plateau (see Volume 1 – Chapters 3 and 4). The strong earthquakes in the region reflect the rupture of the most superficial part of this large thrust, the frontal and middle segments of the fault, locked for centuries during which tectonic stresses build up at mid-crustal depths (see Volume 1 – Chapter 4).

These earthquakes participate in the transfer of tectonic stresses through the fault system, to the partial relaxation of tectonic stresses, and sometimes, for the strongest of them, in the surface expression of plate tectonics. The documentation of these seismic events contributes to varying degrees to the estimation of seismic hazards in the Himalayan region (see Volume 3 – Chapter 6). Their understanding makes it possible to assess the very high seismic hazard, which exposes populations of hundreds of millions of inhabitants living in the countries of the Himalayan arc and its foothills.

5.2.1. *Historical chronicles and earthquakes*

The effects of strong ancient earthquakes are mentioned in the writings of several chroniclers from the northern regions of India and the Himalaya (Iyengar et al. 1999). Several testimonies, sometimes considered as primary sources, sometimes reported in the regional historiography, were preserved. They document a 1,000-year long history of dozens of historical earthquakes.

This is not surprising: the Himalayan range is potentially the site of several independent sources of very strong destructive earthquakes, events that can be felt several hundreds of kilometers from their epicenter. The most important of those events are felt over all or part of the Indian subcontinent (see Figure 8.1 in Volume 3 – Chapter 8).

Figure 5.2. *Historical source of Nepalese medieval earthquakes. Photography of a leaflet of the Gopālarājavaṃśāvalī shows a genealogical record of Nepalese monarchs composed of chronicles compiled in the latest 14th century (Vajrācārya and Malla 1985)*

They are reported on several occasions in medieval chronicles. In Nepal, the Gopālarājavaṃśāvalī manuscript (Figure 5.2, Vajrācārya and Malla 1985) documents the royal chronicles in the Kathmandu valley. It mentions the occurrence of several major seismic events including the devastating earthquakes of 1255 and 1344 (Pant 2002). These two events were each responsible for the death of a king and many inhabitants of the Kathmandu valley (the manuscript even relates the death of a third of the inhabitants in 1255).

Similarly, the 1714 earthquake in western Bhutan was described as having caused widespread destruction in several valleys and is said to have brought down the Gangteng Temple shortly after its completion (Hetényi et al. 2016).

The descriptions of the impact of earthquakes in the few places hosting chroniclers in medieval times inform us about local specific effects of earthquakes that are often difficult to translate in terms of magnitude and location of the seismic sources. Despite these limitations, these historical descriptions are often associated with calendrical dates of the seismic events, and sometimes provide us with facts on the effects of the shaking on the environment or on the duration of the aftershocks, which are important sources of indirect information useful for the characterization of the earthquakes.

5.2.2. *Strong instrumental earthquakes*

Strong Himalayan earthquakes have particularly attracted the attention of scientists since the 19th century, especially since the Nepalese earthquake of 1833, an event that strongly impacted central Nepal and the Kathmandu valley (e.g. Bilham 2019). However, we had to wait until the very end of the 19th century, and the birth of instrumental seismology, to benefit from seismological recordings and extensive documentary bases of the effects of Himalayan earthquakes – sometimes accompanied by photographic bases collected by the teams who reported the damage (Figure 5.3). Our knowledge of the large earthquakes, of their effects and their societal impacts progressed then significantly, thanks to the work of Oldham father and son, of the geological survey of India and of Bhrama Shansher Rana in Nepal, among other contributors. These scientists and their teams documented extensively the damage of the devastating earthquakes of 1897, 1905 and 1934 – and their interpretations in terms of earthquake sources (Figure 5.1) (e.g. Rana 1936; Roy 1939).

Figure 5.3. *Photos of the ruins affecting the historical centre of the Katmandu valley, following two large earthquakes, respectively, on the January 15, 1934 (photo of the Bakhtapur Durbar square from Rana 1936) and on the April 25, 2015. For a color version of this figure, see www.iste.co.uk/cattin/himalaya3.zip*

The testimonies and macroseismic observations collected after the mainshocks, reinterpreted in the light of our modern scientific knowledge, are still a helpful source of information for deciphering the strong earthquakes of the last century.

The early seismological records of the largest Himalayan earthquakes have also been reinterpreted several times in terms of epicentral location, magnitude and sometimes focal mechanisms, as techniques for analyzing seismological waveforms improved (Figure 5.4).

The scarcity of early recordings, essentially available at large (teleseismic) distances, limited our knowledge of these earthquakes to rough main parameters. The confrontation of macroseismic and seismological data with geological field surveys characterizing the tectonic structures activity is therefore crucial for improving our knowledge of the seismic sources. The examination of the geology of the Quaternary is also the only way to access the prehistoric realization of strong earthquakes.

PALEOSISMOLOGY.– What if there is no instrument, building or person to document a seismic event? The scaling laws teach us that there is a proportionality between the magnitude of an earthquake, the dimensions of the associated rupture that propagates along a fault plane and the amplitude of the displacement observed along this rupture.

Magnitude Mw	Rupture length (km)	Maximum displacement (m)
6	10	0.1
7	100	3
8	300	10
9	1,000	30

For strong earthquakes, the rupture width generally exceeds the depth of the hypocenter; the rupture then reaches the surface and leaves a permanent deformation in the landscape. The length of the rupture helps us to determine the magnitude of the earthquake. Over time and successive earthquakes, the deformation accumulates and becomes detectable in the landscape. If the rupture offsets a marker (e.g. an alluvial terrace tread or raiser, a stream, a wall), the displacement is "recorded" in the landscape. By systematically measuring the markers' offset, we can reconstruct the distribution of displacement along the fault and propose an alternative magnitude estimate. Earthquakes occurring on faults, the hypocenter is necessarily located on the rupture plane that emerges on the surface. By determining its dip, we can project it onto the surface and estimate an area in which the epicenter must be located.

Finally, the geologist will naturally turn to stratigraphy to look for a chronology there. By excavating a trench across the surface rupture, the normal and deformed stratigraphic succession reveals deformation episodes (earthquakes) that occurred between depositional episodes. By dating the latter using appropriate methods (according to the available materials and the suspected age), the date of occurrence of some earthquakes can be estimated with a precision of a few decades to a few centuries. By combining geology, geomorphology and seismology, paleoseismology extends our knowledge of strong pre-historic earthquakes and enriches seismicity catalogs with rare and destructive events.

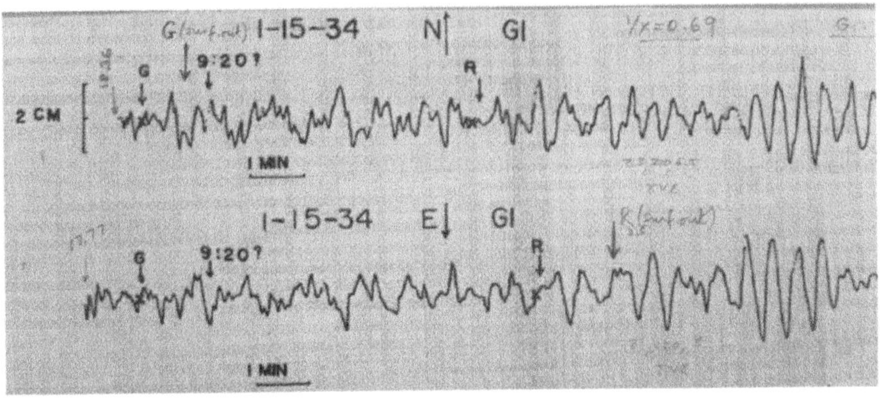

Figure 5.4. *Annotated seismogram of the surface waves of the January 15, 1934 great Bihar–Nepal earthquake recorded on the North and East components of the Wiechert seismograph of the Riverview Sidney observatory (modified from Chen and Molnar (1977)). For a color version of this figure, see www.iste.co.uk/cattin/himalaya3.zip*

5.3. Surface expression of the seismic deformation in the landscape and within paleoseismological excavations

On the morning of October 8, 2005, a very strong earthquake devastated Pakistani Kashmir and in particular the city of Muzaffarabad. This event, of magnitude 7.6, left 80,000 people dead. It is the first of the large Himalayan earthquakes studied jointly by modern techniques of seismology and satellite imagery. These techniques helped to qualify, quantify and extrapolate the local observations of witnesses and scientists mobilized in the field. The observations document a 75-km long surface rupture, accommodating a coseismic slip reaching 7 meters (e.g. Avouac et al. 2006). These observations testify for a surface rupture through the city of Muzaffarabad. This rupture offset outcrops at the base of the topography, and raised a set of alluvial terraces of the Neellum River. Despite the amplitude of the fault ruptures described at the foot of the hangingwall reliefs, the primary surface rupture was often covered by massive landslides.

5.3.1. *In the mesoseismal trace of the 1934 earthquake*

For the inhabitants of central Nepal and the Bihar state in India, the October 2005 Pakistani earthquake evoked the memory of a very strong devastating event, the great Bihar–Nepal earthquake that happened on January 15,

1934. This cataclysmic event, documented by many moving testimonies, is commemorated every year in Nepal on the "National earthquake day".

Largely described in the reports of the Survey of India and in the writings of Brahma Shamsher Rana, the Nepalese General in charge of the 1934 postseismic field surveys in Nepal, the 1934 event macroseismic field appears to be rather complex (Roy 1939; Rana 1936; Martin and Szeliga 2010). Indeed, high earthquake intensities were documented at several places along the so-called "slump belt" , (a) a region in the state of Bihar exposed to massive liquefactions and regional subsidence, (b) in Kathmandu valley, as well as (c) in southeastern Nepal, where massive landslides were mapped. Despite the extensive surveys, the field campaigns carried out after the earthquake did not allow observers at the time to highlight any surface rupture. The geological signature of this event therefore remained unknown for decades.

Field campaigns carried out along the Himalayan frontal thrust in the mesoseismal zone of the Bihar–Nepal earthquake revealed ruiniform landforms and seismic scarps. Some of the most spectacular escarpments are several tens of meters-high – corresponding to cumulative multiple events seismic scarps – and sometimes well exposed after being cut by rivers (Sapkota 2011). This is the case of the 31 meters high scarp stripped by the Sir Khola River (Figure 5.5). This outcrop reveals several faults, outlined by dark gouges, which affect Siwaliks siltstones (see Volume 2 – Chapter 8 for the geology of the foreland basin). These rocks that date back from Mio-pliocene times are locally thrusted over Quaternary alluvials, and recent deposits of the Sir Khola river, testifying for recent tectonic movements (Sapkota et al. 2013).

The examination of the most frontal part of the outcrop reveals that the gravels and pebbles of the river alluvials are drag-folded beneath the fault (Figure 5.6). The excavation of a 43-meter long 5-meter deep paleoseismic trench, eastward, at the toe of the cumulated scarp reveals additional tectonic expressions and relations between the faulted Siwalik rocks and several units of collapse wedges and alluvial deposits.

The detailed logging of these units and the radiocarbon dating of 28 samples help constrain the recent history of the outcrop. The youngest sediments found affected by deformation in the footwall of the fault date back from 1660 to 1950 AD. This demonstrates that the latest rupture on the frontal fault happened sometimes between the second half of the 17th century and the

1950s, a period during which the historical archives comprise only one major earthquake accompanied locally with intensities I $\geq$ VIII, the 1934 earthquake.

This event is therefore the only candidate known so far by historians to rupture the frontal faults at Sir Khola.

Figure 5.5. *Annotated photo of the cumulated seismic scarp at Sir Khola river in Nepal. The rivercut cliff exposes faults, some of them outlined by dark gouges. Som Nath Sapkota, standing on the floodplain gravels, takes a picture of Yann Klinger, standing on Fault F4. The white rectangle locates the frame of Figure 5.6. For a color version of this figure, see www.iste.co.uk/cattin/himalaya3.zip*

Similar remnants of recent ruptures were excavated further east, along two rivers, in the mesoseismal area of the 1934 earthquake (e.g. Rizza et al. 2019).

The radiocarbon dating of the most recent sedimentary units affected by the deformation demonstrates that the latest earthquake postdates the early 17th century and predates the 1950s, demonstrating that the 1934 Bihar–Nepal earthquake rupture was not blind.

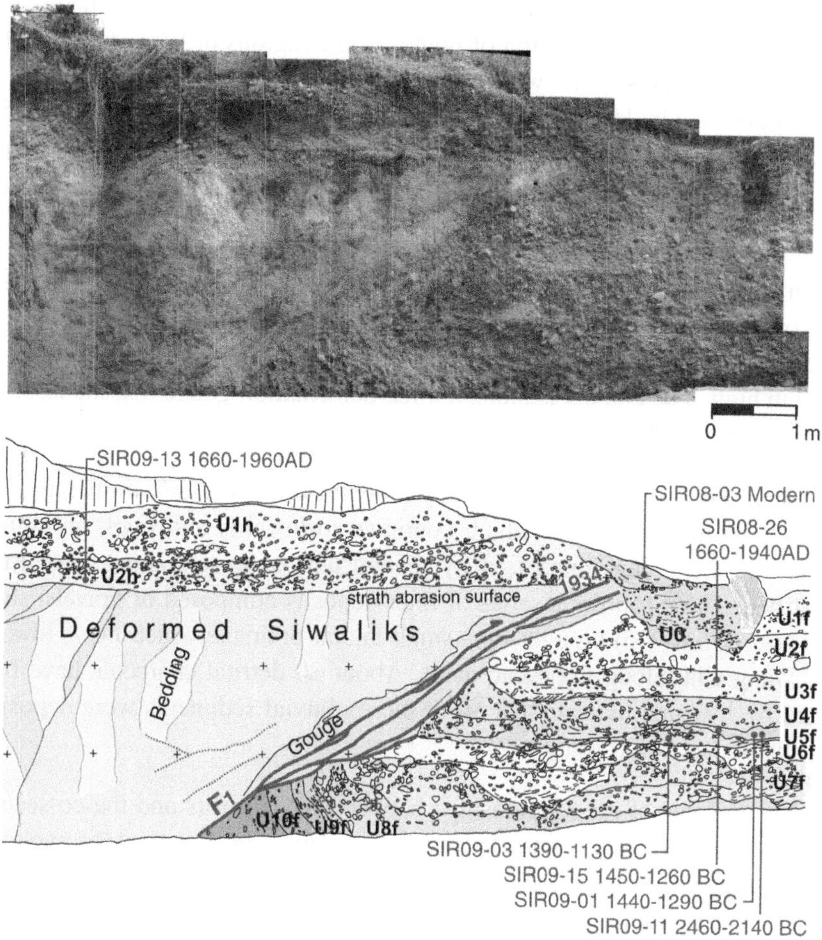

Figure 5.6. *Photomosaic close up view on fault F1, a 27° N dipping thrust fault emplacing the Siwaliks siltstones (light beige) on top of the light brown gravels-pebbles of alluvial sediments, drag folded beneath the fault. In the hangingwall, the Siwaliks are covered by gravels and a colluvial wedge deposited at the toe of the cumulative scarp (modified from Sapkota et al. (2013)). For a color version of this figure, see www.iste.co.uk/cattin/himalaya3.zip*

5.3.2. *In the mesoseismal trace of the 1714 earthquake in Bhutan*

Although quite recent, the 1714 earthquake is not well known because there are only a few historical texts covering this event. Most of the descriptions came to us thanks to the testimony of Shakya Rinchen, a 4-year-old child who would later become a Buddhist monk and then spiritual leader of the country. He reports massive destruction of homes in the regions of Wangdue, Punakha and Thimphu. A second witness, a famous temple builder, lamented the destruction of a building just completed in Gangteng. The damage distribution seems to indicate a meso-seismal zone covering central and western Bhutan, associated with a maximum intensity between VIII and IX (Hetényi et al. 2016).

In this region, the MFT emerges on the banks of the Wang Chu River and offers a spectacular outcrop where the Siwalik conglomerates can be seen overlapping a Holocene alluvial series (Figure 5.7) forming an escarpment of ~30 m high. The contact is along a thick deformation gouge that incorporates elements from adjacent units.

Here, the Siwaliks are covered by a coarse alluvial unit with pebbles and blocks (U7 in Figure 5.7) which continue northwards and mark an alluvial terrace of the Wang Chu raised by several tens of meters above the current river bed. There follows a series of finer deposits composed of gravels, sands and silts (U6 to U0) brought by a small lateral drainage which flows towards the Wang Chu along the escarpment. About 20 detrital charcoals have been dated by ^{14}C and indicate that all of these alluvial sediments were deposited within the last 3,000 years.

The analysis of the relationships between the deposits and the co-seismic deformations identified on the log made it possible to reconstruct the evolution of the escarpment over time. Thus, Figure 5.8, although schematic, illustrates the chronology of the deposits and the ruptures that affect them, and it serves as a basis for determining the ages of the paleoearthquakes and their associated displacements.

The most recent earthquake (event 1) visible in the highest part (sub-vertical F5 rupture) is later than the year 895 and offset the superficial units by ~1.5 m, two characteristics consistent with the date and the estimated magnitude from the 1714 earthquake. Very similar observations were also noted ~50 km further east in trenches excavated across the MFT at Sarpang and Gelephu

(Le Roux-Mallouf et al. 2016). It is on the basis of these observations combined with historical testimonies that a modeling of the intensity of the earthquake confirms a very probable location in western Bhutan and specifies its magnitude at Mw 8.0 ± 0.5 (Hetényi et al. 2016): the earthquake of 1714 would have therefore a well broken surface over a length of ~200 km.

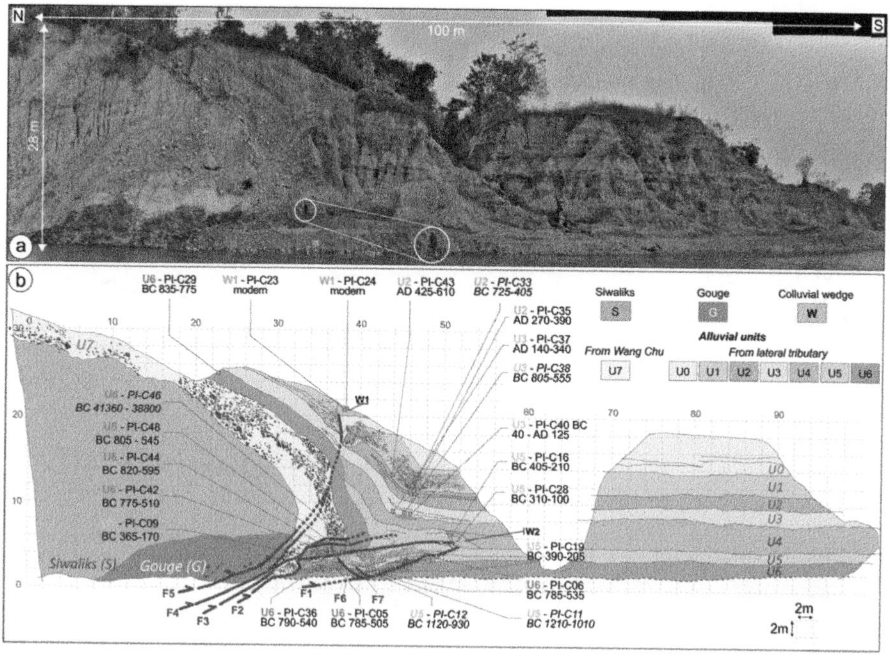

Figure 5.7. *The Piping outcrop (western Bhutan) relates the emergence of the Siwaliks unit (denoted S) along the MFT (faults denoted F1 to F7) and the deformation of a sedimentary pile deposited within the last three millennia (strata denoted U7 to U0). (a) Panoramic orthophoto showing the extent of the site and the main units. Note the person in the circle for scale. (b) Corresponding stratigraphic log accompanied by ^{14}C dating (modified from Le Roux-Mallouf et al. 2016). For a color version of this figure, see www.iste.co.uk/cattin/himalaya3.zip*

Lower in the stratigraphy, four other earthquakes deform the older layers with much larger co-seismic displacements, of the order of 10–15 m, which opens the door to even stronger magnitudes which approach or even exceed Mw 9 and surface failure lengths exceeding 400 km.

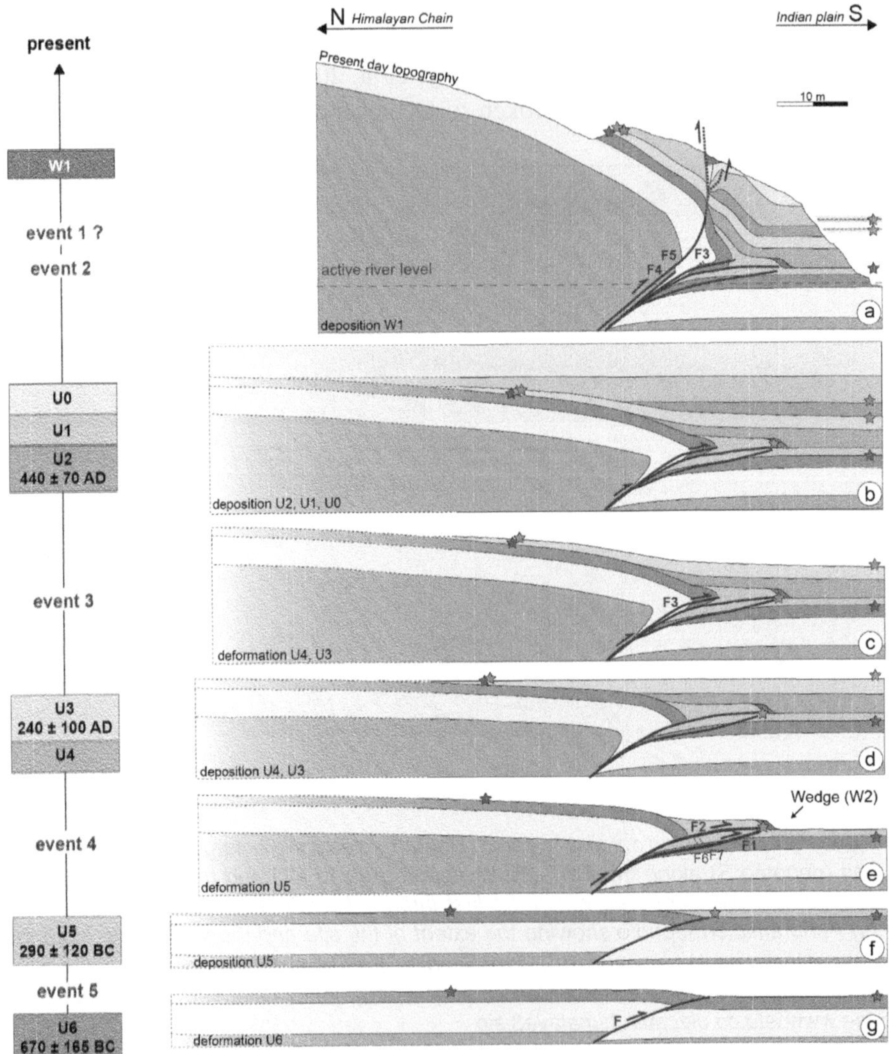

Figure 5.8. *Schematic reconstruction of the episodes of deposition (colored units) and deformation (events) that led to the current situation (diagram a). The dating by ^{14}C of around 20 pieces of detrital charcoal sampled in the fine deposits makes it possible to establish a chronostratigraphic framework and to deduce the approximate date of each earthquake (modified from Le Roux-Mallouf et al. 2016). For a color version of this figure, see www.iste.co.uk/cattin/himalaya3.zip*

5.3.3. *In the mesoseismal trace of the 1950 earthquake*

Despite being one of the largest recent continental earthquakes, the magnitude Mw 8.7 earthquake of August 15, 1950 remains poorly studied. Its source, mechanism and surface trace are still subject of discussion (Jayangondaperumal et al. 2011; Priyanka et al. 2017; Coudurier-Curveur et al. 2020).

A recent reappraisal of available seismological and geological data, together with field observations and measurements of surface deformation, allow for a better description and understanding of this large event (Coudurier-Curveur et al. 2020). Only recently, in the last decade, clearance of the densely forested foothills of the eastern Himalaya (Abor and Mishmi hills) due to agricultural pressure has unveiled the presence of a continuous over 300-km long tectonic scarp partly rejuvenated during the 1950 event.

A major and sensible consequence of this major earthquake in the steep forested mountains of eastern Assam has been the many landslides that affected the shaken area (see Volume 3 – Chapter 4). A witness of the event near the epicentral area remembers "*it was difficult to separate the noise made by the earthquake itself from the roar of the rock avalanches pouring down on all sides*" (Kingdon-Ward 1951). The many rock avalanches triggered by the earthquake covered the region in a fog of dust for weeks and exposed the bare bedrock in many of the steep slopes of the Abor and Mishmi hills, some of the scars are still visible today. The mud and debris now filling the valleys were soon to be transported into the Assam valley, filling the beds of the large rivers that started flooding, an event that will recur several years after the earthquake. The tree trunks that were progressively washed away from the foothills make up a specific layer that can be seen today in the lower terraces of the large rivers such as the Brahmaputra and Lohit.

From Subansiri in the west to Vijaynagar, near the boundary of Myanmar in the east, a several meter high scarp (5–14 m) can be followed almost continuously for more than 250 km, only vanishing in the active bouldery riverbeds of the large rivers of eastern Assam. Aftershock distribution and depth, local field observation of the thrusts in river sections, relief differences between the Abor and Mishmi hills, all concur to suggest a shallow thrust below the western Abor Hills section (dip of 10–20°N) and a steep thrust below the eastern Mishmi Hills section (30° E) (Coudurier-Curveur et al. 2020). In the west a co-seismic vertical increment of a few meters in the

1950 event contributes to the 5-meter high scarp like in Niglok, while about 7 m are believed to be accumulated during that same event along the Mishmi Hills (Figures 5.9 and 5.10). Such differences in vertical accumulation are compatible with 10–20 m of slip at depth on the thrusts with distinct dips. The age range of terraces offset 14 m vertically, that is, by two 1950-type events, suggests that such an event would recur every 1,200–2,600 years.

Figure 5.9. *View to north of Tebang river cut west northwest of Lohitpur along Mishmi Hills in India. East dipping gray conglomerate overthrusts along shallow 20° N dipping reverse fault, orange sub-horizontal to overturned modern conglomerate, also visible atop the 40-m high terrace wall. For a color version of this figure, see www.iste.co.uk/cattin/himalaya3.zip*

Figure 5.10. *View towards northeast of 5–7 m-high scarp of Abor Hills Main Frontal Thrust southwestward of Pasighat. For a color version of this figure, see www.iste.co.uk/cattin/himalaya3.zip*

5.4. Overview of the paleoseismological excavations along the Himalayan arc

Extensive field surveys along the meso-seismal traces of the strongest instrumental and historical earthquakes helped discover remnants of their surface ruptures. Since then, more than 27 paleoseismological trench sites have been excavated along the Himalayan arc (Figure 5.11) (e.g. Nakata 1989; Lavé et al. 2005; Kumar et al. 2006; Kondo et al. 2008; Jayangondaperumal et al. 2011; Mugnier et al. 2013; Rajendran et al. 2013; Sapkota et al. 2013; Bollinger et al. 2014; Malik et al. 2015; Le Roux-Mallouf et al. 2016; Wesnousky et al. 2017; Rizza et al. 2019; Coudurier-Curveur et al. 2020). Radiocarbon dating of the charcoals sampled in these excavations revealed the age of the latest surface rupturing earthquakes. With the exception of the "modern" ages determined along the trace of the 1950 earthquake, the ages of the latest earthquakes correspond either to medieval ages or to undetermined ages between the 17th century and the middle of the 20th century.

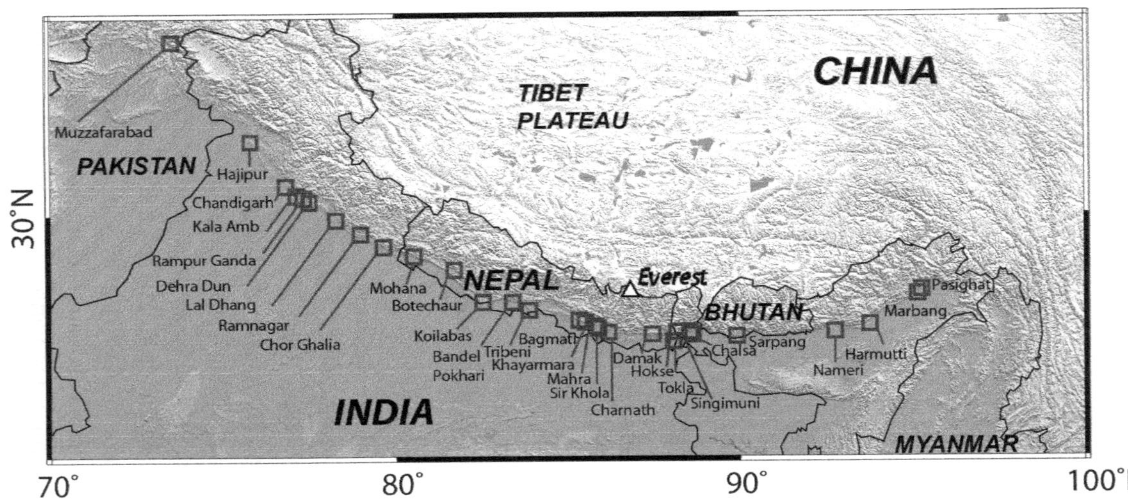

Figure 5.11. *Paleoseismological sites excavated along the Himalayan front (e.g. (e.g. Nakata 1989; Lavé et al. 2005; Kumar et al. 2006; Kondo et al. 2008; Jayangondaperumal et al. 2011; Mugnier et al. 2013; Rajendran et al. 2013; Sapkota et al. 2013; Bollinger et al. 2014; Malik et al. 2015; Le Roux-Mallouf et al. 2016; Wesnousky et al. 2017; Coudurier-Curveur et al. 2020)). For a color version of this figure, see www.iste.co.uk/cattin/himalaya3.zip*

These ages are sometimes accompanied by large uncertainties resulting from: (1) the analytical and calibration uncertainties associated with the radiocarbon dating of detrital charcoals – the most widely used method for dating Himalayan paleoearthquakes – (2) biases induced by the sampling and indirect dating of sediment deposition?, that is, the age of the detrital charcoals is older than the deposition of the sediments in which it was trapped (3) as well as a high epistemic uncertainty, related to the discontinuity of the sedimentary record or absence of dating material.

Despite these challenges, several paleoearthquakes ruptures excavated along the front of the Himalaya were associated with historical earthquakes, documented by the local chronicles. However, the calendar ages of a few ruptures dating back from medieval times, in region depleted in historical macroseismic observations, are still debated.

Another challenge for the paleoseismologists in Himalaya is that only very rare trenches reveal the ruptures of two or more earthquakes. The excavations of the last 20 years therefore do not yet make it possible to resolve the average return time of earthquakes along most segments of the front, on the sole basis of paleoseismic trenches.

The return period of large earthquakes could be roughly estimated from the confrontation of the geodetic slip rate on the fault and the coseismic offsets measured in the trench. However, the evaluation of accurate coseismic offsets remains usually challenging, as the trenches exposing the deformation accommodated through a few faults spread over a few tens of meters while the coseismic slip is probably transferred through multiple fault zones, over a width of dozens to hundreds of meters.

5.5. Abandoned alluvial terraces, an archive of the paleoearthquakes

Large Himalayan earthquakes are followed by significant disturbances affecting the course of rivers and the sedimentary dynamics in general. Indeed, the frontal hills, all along the surface rupture, are affected by massive landslides. The sediments mobilized after the earthquake are evacuated by the rivers that transport heavy loads towards the foreland, raising in the meantime their abrasion potential. The course of these rivers and their capacity to incise into their riverbed are also locally affected by the uplift of the hangingwall of the faults that were activated, and the formation of dams and/or rapids at the seismic scarps.

Mapping the river terraces and dating their deposits and abandonment make it possible to constrain their ancient history, which comprise their response to local earthquakes. Strong earthquakes are not the only events to be associated with episodes of massive transport, deposition and incision, eventually accompanied by permanent abandonment of alluvial terraces. Indeed, massive landslides or climatic episodes can induce a significant alluvial/sedimentary response. But the signature of tectonic events presents some differences, in terms of spatial imprint or time evolution. The tectonic abandonments of terraces have been recognized at various places where surface ruptures of Himalayan earthquakes were dated.

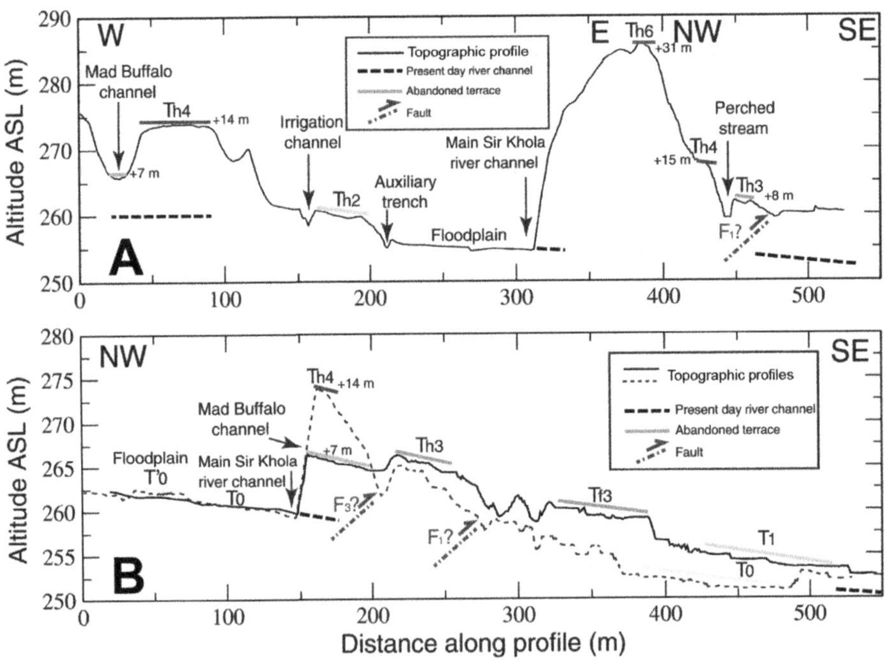

Figure 5.12. *Cross-sections through the topography and abandoned alluvial terraces along the Sir Khola river (modified from Bollinger et al. (2014)). For a color version of this figure, see www.iste.co.uk/cattin/himalaya3.zip*

More than six generations of alluvial terraces, suspected to have been abandoned after tectonic events, were dated in eastern Nepal documenting a 5,000-year long history of earthquakes (Figure 5.12). The average return period of surface rupturing earthquakes deduced from that work falls between 750 ± 140 and 870 ± 350 years (Bollinger et al. 2014).

At some places, the amount of incision associated with a tectonic event can lead to the estimation of an "apparent uplift", when the terrace abandonment by the river is followed by a rapid return to its equilibrium. This apparent uplift is then translated into an apparent coseismic slip when the geometry of the fault is documented at depth, below the alluvial terraces (see Volume 1 – Chapter 7). In those cases, mapping and dating alluvial terraces thus appears to be a powerful tool, complementary to the excavation of the most superficial ruptures.

5.6. Conclusion

Fieldworks carried out over the past 20 years along the Himalayan front associated with observations of the surface rupture of the 2005 Muzzafarabad earthquake in Pakistan helped improve our capacity to read in the Himalayan landforms and reveal the geological signature of ancient earthquakes.

Remnants of several of the great historical earthquakes that happened along the Himalayan range were found in the field, along rivercut outcrops and/or within trenches excavated at the toe of the most frontal scarps. The very large size of surface coseismic offsets induced by these earthquakes sometimes exceeds 10 meters. Despite these large amounts of displacements and the gigantisms of the seismic scarps found along the frontal faults and folds, the great earthquakes are followed by hydrogeological episodes that often hide or erase the surface signature of the earthquakes. These processes are often induced by landsliding, massive sediment transport by the rivers, sediment aggradation and/or sharp incision and abandonment of the river courses that follow the earthquakes realization. All this makes it difficult to retrieve multiple seismic events in a single trench. Dating the remnants of the main geomorphic episodes that affect the rivers that flow through the frontal faults appears to be interesting for complementing the paleoearthquake records, and helpful for determining past earthquakes lateral extents.

5.7. References

Avouac, J.P., Ayoub, F., Leprince, S., Konca, O., Helmberger, D.V. (2006). The 2005, Mw 7.6 Kashmir earthquake: Sub-pixel correlation of ASTER images and seismic waveforms analysis. *Earth and Planetary Science Letters*, 249(3–4), 514–528.

Bilham, R. (2019). Himalayan earthquakes: A review of historical seismicity and early 21st century slip potential. *Geological Society, London, Special Publications*, 483(1), 423–482.

Bollinger, L., Sapkota, S.N., Tapponnier, P., Klinger, Y., Rizza, M., Van Der Woerd, J., Tiwari, D.R., Pandey, R., Bitri, A., Bes de Berc, S. (2014). Estimating the return times of great Himalayan earthquakes in eastern Nepal: Evidence from the Patu and Bardibas strands of the Main Frontal Thrust. *Journal of Geophysical Research*, 119, 7123–7163

Chen, W.P. and Molnar, P. (1977). Seismic moments of major earthquakes and the average rate of slip in central Asia. *Journal of Geophysical Research*, 82(20), 2945–2969.

Coudurier-Curveur, A., Tapponnier, P., Okal, E., Van der Woerd, J., Kali, E., Choudhury, S., Baruah, S., Etchebes, M., Karakaş, Ç. (2020). A composite rupture model for the great 1950 Assam earthquake across the cusp of the East Himalayan Syntaxis. *Earth and Planetary Science Letters*, 531, 115928.

Hetényi, G., Le Roux-Mallouf, R., Berthet, T., Cattin, R., Cauzzi, C., Phuntsho, K., Grolimund, R. (2016). Joint approach combining damage and paleoseismology observations constrains the 1714 AD Bhutan earthquake at magnitude 8±0.5. *Geophysical Research Letters*, 43(20), 10-695.

Iyengar, R.N., Sharma, D., Siddiqui, J.M. (1999). Earthquake history of India in medieval times. *Indian Journal of History of Science*, 34, 181–238.

Jayangondaperumal, R., Wesnousky, S.G., Choudhuri, B.K. (2011). Near-surface expression of early to Late Holocene displacement along the northeastern Himalayan frontal thrust at Marbang Korong Creek, Arunachal Pradesh, India. *Bulletin of the Seismological Society of America*, 101(6), 3060–3064.

Kingdon-Ward, F. (1951). Notes on the Assam earthquake. *Nature*, 167, 130–131. https://doi.org/10.1038/167130a0.

Kondo, H., Nakata, T., Akhtar, S.S., Wesnousky, S.G., Sugito, N., Kaneda, H., Tsutsumi, H., Khan, A.M., Khattak, W., Kausar, A.B. (2008). Long recurrence interval of faulting beyond the 2005 Kashmir earthquake around the northwestern margin of the Indo-Asian collision zone. *Geology*, 36(9), 731–734.

Kumar, S., Wesnousky, S.G., Rockwell, T.K., Briggs, R.W., Thakur, V.C., Jayangondaperumal, R. (2006). Paleoseismic evidence of great surface rupture earthquakes along the Indian Himalaya. *Journal of Geophysical Research*, 111, B03304.

Lavé, J., Yule, D., Sapkota, S., Basant, K., Madden, C., Attal, M., Pandey, R. (2005). Evidence for a great Medieval earthquake (~1100 AD) in the central Himalayas, Nepal. *Science*, 307, 1302–1305.

Le Roux-Mallouf, R., Ferry, M., Ritz, J.F., Berthet, T., Cattin, R., Drukpa, D. (2016). First paleoseismic evidence for great surface-rupturing earthquakes in the Bhutan Himalayas. *Journal of Geophysical Research: Solid Earth*, 121(10), 7271–7283.

Malik, J.N., Sahoo, S., Satuluri, S., Okumura, K. (2015). Active fault and paleoseismic studies in Kangra valley: Evidence of surface rupture of a great Himalayan 1905 Kangra Earthquake (Mw 7.8), Northwest Himalaya, India. *Bulletin of the Seismological Society of America*, 105, 2325–2342.

Martin, S. and Szeliga, W. (2010). A catalog of felt intensity data for 570 earthquakes in India from 1636 to 2009. *Bulletin of the Seismological Society of America*, 100, 562–569.

Mugnier, J.L., Gajurel, A.P., Huyghe, P., Jayangondaperumal, R., Jouanne, F., Upreti, B. (2013). Structural interpretation of the great earthquakes of the last millennium in the central Himalaya. *Earth-Science Reviews*, 127, 30–47.

Nakata, T. (1989). Active faults of the Himalaya of India and Nepal. *Geological Society of America Special Papers*, 232, 243–264.

Pant, M.R. (2002). A step toward a historical seismicity of Nepal. *Adarsa*, 2, 29–60.

Priyanka, R.S., Jayangondaperumal, R., Pandey, A., Mishra, R.L., Singh, I., Bhushan, R., Srivastava, P., Ramachandran, S., Shah, C., Kedia, S. et al. (2017). Primary surface rupture of the 1950 Tibet-Assam great earthquake along the eastern Himalayan front, India. *Science Reports*, 7(5433). https://doi.org/10.1038/s41598-017-05644-y.

Rajendran, C.P., Rajendran, K., Sanwal, J., Sandiford, M. (2013). Archeological and historical database on the medieval earthquakes of the central Himalaya: Ambiguities and inferences. *Seismological Research Letters*, 84, 1098–1108.

Rana, B.S. (1936). Mahabhukampa, Bahar Mahal BS1992. Report, Kathmandu.

Rizza, M., Bollinger, L., Sapkota, S.N., Tapponnier, P., Klinger, Y., Karaka, C., Kali, E., Etchebes, M., Tiwari, D.R., Siwakoti, I. et al. (2019). Post-earthquake aggradation processes to hide surface ruptures in thrust systems: The M8.3, 1934, Bihar-Nepal earthquake ruptures at Charnath Khola (Eastern Nepal). *Journal of Geophysical Research*. https://doi.org/10.1029/2018JB016376.

Roy, S.C. (1939) The Bihar-Nepal earthquake of 1934. In *Memoirs of the Geological Survey of India*, 73. Government of India, Kolkata.

Sapkota, S.N. (2011). Surface rupture of 1934 Bihar-Nepal earthquake: Implications for seismic hazard in Nepal Himalaya. Doctoral Dissertation, Institut de physique du globe, Paris.

Sapkota, S.N., Bollinger, L., Klinger, Y., Tapponnier, P., Gaudemer, Y., Tiwari, D. (2013). Primary surface ruptures of the great Himalayan earthquakes in 1934 and 1255. *Nature Geoscience*, 6, 71–76.

Vajrācārya, D. and Malla, K.P. (1985). *The Gopālarājavaṃśāvalī*, Vol. 9. Franz Steiner Verlag Wiesbaden GmbH.

Wesnousky, S.G., Kumahara, Y., Chamlagain, D., Pierce, I.K., Karki, A., Gautam, D. (2017). Geological observations on large earthquakes along the Himalayan frontal fault near Kathmandu, Nepal. *Earth and Planetary Science Letters*, 457, 366–375.

6

Seismic Coupling and Hazard Assessment of the Himalaya

Sylvain MICHEL[1], Victoria STEVENS[2,3], Luca DAL ZILIO[4] and Romain JOLIVET[1]

[1] *Ecole Normale Supérieure, Paris, France*

[2] *Geological Sciences Department, University of Cape Town, Rondebosch, South Africa*

[3] *Nanyang Technological University, Singapore*

[4] *ETH Zürich, Switzerland*

6.1. Introduction

The Himalaya, the highest mountain range in the world, bounds one of the most populated regions on Earth, extending from northern Pakistan to northern India and Bangladesh across Nepal and Bhutan, leaving millions of people under the threat of major earthquakes. Seismic hazard assessments have important implications for society, providing a basis for building codes and public policies (e.g. Shaw et al. 2018). Such assessments require a careful evaluation of the history of past earthquakes as well as a profound understanding of active tectonic structures and current surface strain rates. In this chapter, we provide a description of what is currently known in

Himalaya, Dynamics of a Giant 3,
coordinated by Rodolphe CATTIN and Jean-Luc EPARD.

the Himalaya about regional and local seismicity, the surface strain budget, the seismically active structures and how to incorporate these elements into seismic hazard assessment.

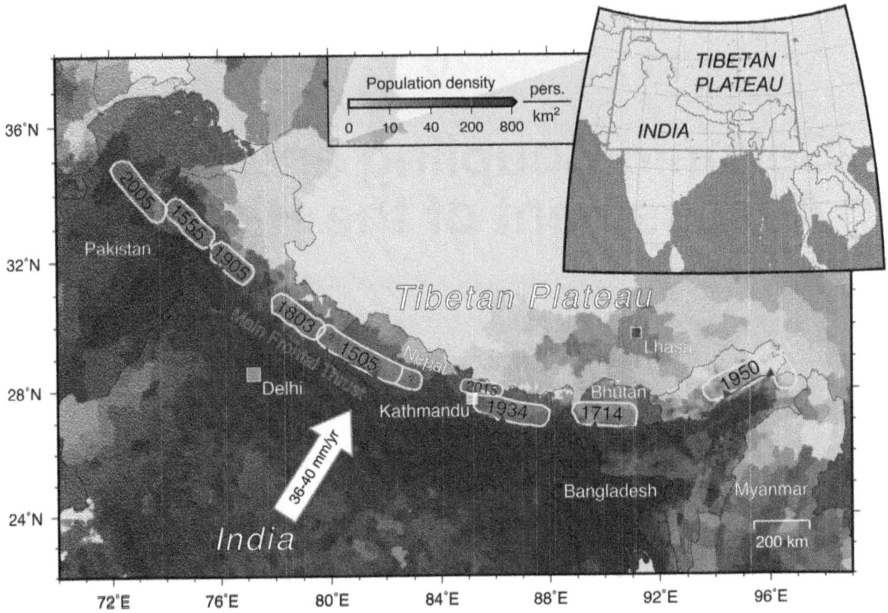

Figure 6.1. *Past earthquakes and population density in the Himalaya. Yellow shapes delimit the extent of past ruptures for events of magnitude>7.5 published in the literature (Bilham 2019; Bilham and England 2001; Bollinger et al. 2014; Kumar et al. 2010). The blue background is the population density (CIESIN 2009). White arrow indicates the direction of motion of the Indian plate with respect to Eurasia. Only 20 mm/yr are accommodated by the Main Himalayan Thrust, indicated as a red line (modified from Hetényi et al. (2016) and Dal Zilio et al. (2020)). For a color version of this figure, see www.iste.co.uk/cattin/himalaya3.zip*

Over the past hundreds of years, the Himalaya has been struck by multiple large earthquakes (Figure 6.1). The most recent is the 2015 Mw 7.8 Gorkha earthquake which rattled the city of Kathmandu (e.g. Avouac et al. 2015; Galetzka et al. 2015; Elliott et al. 2016, see also Volume 3 – Chapter 8). Previously, the chain witnessed even larger events, such as the Mw 8.5 Assam earthquake in 1950 or the Mw 8.2 Nepal–Bihar earthquake in 1934 (Chen and Molnar 1990; Sapkota et al. 2013; Bilham 2019; Dal Zilio et al. 2021). A summary of past ruptures can be found in Bilham (2019), but it is to be noted that all earthquakes that pre-date the 20th century were inferred either from indirect testimonies or from paleo-seismological investigations leading to uncertain rupture extents, a key ingredient for future scenarios of earthquake ruptures (see Volume 3 – Chapter 5). One particular area of concern is the large segment lying to the west of Kathmandu where the last major event occurred in 1505, leaving a large seismic gap which has been accumulating tectonic stress since then.

All these large magnitude events occurred along the Main Himalayan Thrust (MHT), the main thrust system separating the Indian plate from the orogenic wedge (see Volume 1 – Chapters 3, 4 and 6). This major plate boundary slides episodically during earthquakes and on the long-term accommodates about 2 cm/yr of compression between India and southern Tibet. Over long periods of time, typically on the order of 100s of years, the MHT is kinematically locked while the Indian plate is continuously subducting beneath Tibet. As a result, elastic stress increases while the slip deficit builds up and the corresponding elastic energy stored in the crust will be released during future major earthquakes. During this interseismic period, the combination of locking along the main plate boundary and continuous motion of tectonic plates leads to surface strain that can be interpreted in terms of elastic, recoverable deformation of the crust. Analyzing such deformation and the ongoing micro- and moderate seismic activity should therefore provide hints about the amount of elastic energy available for future large earthquakes on the MHT, a key ingredient for seismic hazard assessment in the Himalaya.

However, within the orogenic wedge above the MHT, numerous active faults accommodate part of the deformation, including, for instance, convergence parallel normal faults, accommodating extension due to the geometry of the collision (e.g. Armijo et al. 1986; Copley et al. 2011). These faults and the associated hazard are difficult to analyze especially since evidence for local albeit destructive earthquakes in the past is scarce (e.g. Ghazoui et al. 2019). These structures must be accounted for as they correspond to a distinct threat for the local population.

Evaluating the seismic hazard in a region typically involves finding the probabilities of different earthquakes occurring, including their magnitudes and locations, and then estimating the amount of ground shaking at different locations caused by these earthquakes. Such an approach, termed Probabilistic

Seismic Hazard Assessment (PSHA), involves (1) identifying the source of potential future earthquakes, (2) defining/inferring the distribution of magnitudes of potential earthquakes and the probability of occurrence of each magnitude, (3) choosing ground motion prediction equations and (4) deriving maps of seismic hazard. It is important to note that in this chapter we will not address secondary hazards such as landslides or rockfalls, glacial outburst floods, liquefaction or avalanches that may all be triggered by earthquakes (see instead Volume 3 – Chapters 3 and 4). To perform PSHA in the Himalaya, we must address a few important questions, and we will summarize some key findings before describing how these can be turned into a comprehensive seismic hazard assessment.

First, we will describe current surface strain rate and how such surface deformation informs us on the state of the main faults that might rupture in the future. We will consider surface deformation rates as measured by GNSS (Global Navigation Satellite System) at instrumented sites over the whole Himalayan range and derive a probabilistic model of coupling (i.e. kinematic locking) of the MHT. Second, we will turn this model into a budget of slip along the main plate boundary in order to derive a meaningful probabilistic assessment of earthquake occurrence per magnitude range. The latest models now include some elements of earthquake physics, and we will emphasize the importance of fundamental studies of fault mechanics to shape the probability of occurrence of earthquakes in the region. Third, we will illustrate how to use these estimates to derive seismic hazard maps, turning probabilities of occurrence of earthquakes into probabilities of ground shaking over the next decades. Most of the methods and results discussed here are described in details by Stevens et al. (2018), Dal Zilio et al. (2020), Michel et al. (2021), and Stevens and Avouac (2021).

6.2. From current ground motion to the buildup of slip deficit at depth

Over geological timescales (10^4–10^7 years), crustal deformation is usually described as a continuous, quasi-steady process accommodated by a combination of viscous deformation at depth and brittle deformation in the upper crust (Dal Zilio et al. 2021). However, at shorter timescales ($<10^4$ years), tectonic shortening imposed by plate motion on locked, seismogenic plate boundary faults leads to elastic stress increase, which is ultimately released by earthquakes (e.g. Avouac et al. 2015; Lay 2015). Thus, the occurrence of strong (6<Mw<7), major (7<Mw<8) and great (Mw 8+) earthquakes attests to the episodic nature of upper crustal shortening and associated transient processes (Dal Zilio et al. 2021). Geodetic observations allow us to derive spatial variations of strain rates induced by interseismic locking of the main boundary faults. As plates are moving continuously, the locking of the plate interface leads to the elastic bending of the lithosphere and the subsequent buildup of elastic energy, which can be interpreted as a slip deficit on the said plate boundary. Inferring such a locking distribution amounts to mapping which portions of a plate boundary fault are locked versus those that slip aseismically (e.g. Avouac (2015b), for a review). In the Himalaya, as most large earthquakes have ruptured the Main Himalayan Thrust, we seek to identify locked and slipping portions of this main plate boundary between the Indian and Eurasian plates.

Fault locking is referred to as coupling and is defined as the ratio of the deficit of slip rate in the interseismic period divided by the long-term slip rate imposed by tectonic shortening. This ratio varies from 0 for a portion of the MHT that slips aseismically at a rate equal to the plate convergence rate to 1 for a fully locked fault (Savage 1983). Coupling is therefore a kinematic quantity and should only be interpreted as such. Many models derived from geodetic data suggest interseismic coupling on subduction megathrusts worldwide is spatially heterogeneous, defining locked patches where stress increases more rapidly surrounded by segments that slip aseismically (e.g. Avouac 2015b; Métois et al. 2016). Large megathrust earthquakes occur preferentially on these locked patches, which are prone to weaken dramatically and slip rapidly with the associated radiation of seismic waves.

Most published models in the Himalayan region indicate that the entire MHT is uniformly locked from the surface to beneath the front of the high range over a width of 100±20 km (Ader et al. 2012; Stevens and Avouac 2015). Such a distribution is justified by the fact that GNSS displacement rates are all roughly parallel to each other, pointing to no significant variations in fault locking. Such continuous locking would make the MHT an anomaly with respect to subduction megathrusts worldwide where coupling is highly variable. However, a fully locked MHT does not explain the variable along-strike size and location of large (M7+) Himalayan earthquakes, which suggests segmentation at some level (Figure 6.1). A recent and detailed GNSS velocity solution over Bhutan highlights local variations in coupling over a small fraction of the MHT, with aseismic slip at shallow depth (Maréchal et al. 2016). Micro and moderate seismicity throughout the Himalayan arc varies significantly with some well-defined active regions (Arora et al. 2012, and Volume 1 – Chapter 4). This lateral variation in seismic productivity may result from variations in fault geometry, rheology, frictional properties along the fault interface or structural complexities (Cattin and Avouac 2000; Hubbard et al. 2016; Bilham et al. 2017; Mugnier et al. 2017; Dal Zilio et al. 2019). It is therefore important to evaluate the robustness of such a continuous coupling distribution.

6.2.1. *Geodetic observations of the interseismic period*

The latest assessments of interseismic coupling are based on an updated set of GNSS-derived velocities spanning the 1990s to 2015 (Lindsey et al. 2018, and Figure 6.2). To better resolve the pattern of coupling, we also include leveling measurements from the Survey of Nepal collected between 1977 and 1990 (Jackson and Bilham 1994). Leveling data are traditional topographic measurements which, repeated over time, provide measurements of interseismic uplift rate. Uncertainties on the leveling data are low with respect to those on the GNSS-derived vertical rates and help to constrain the pattern of coupling locally (Ader et al. 2012).

6.2.2. *Inferring coupling along the MHT with a Bayesian analysis*

Our latest fault model considers the total length of the arc of roughly 2,000 km, from 73°E to 96°E (Figure 6.2), with a dip angle set to 10° (Ader et al. 2012; Stevens and Avouac 2015). Due to the arcuate and irregular shape, we divide the fault geometry into four planar sub-faults blocks. These sub-faults are discretized into triangles of variable size, smaller near the surface and larger down-dip. On each of these triangles, we solve for the best distribution of fault locking that will fit the GNSS and leveling data. For details of the modeling approach, we adopt the backslip formulation of Savage (1983) and compute the elasto-static Green's functions in a layered elastic half-space based on the elastic properties of the region (Pandey et al. 1995).

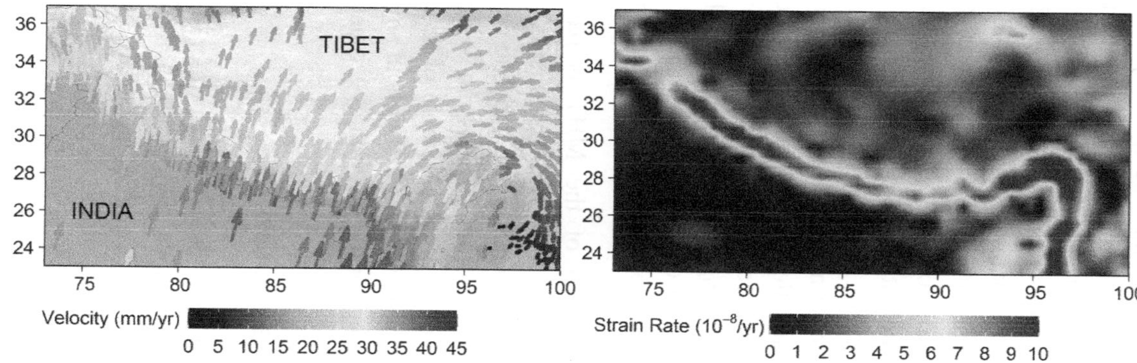

Figure 6.2. *Arrows to the left are the ground velocities measured at GNSS sites over the whole plate boundary, extending from India to Tibet. Color indicates the amplitude of each velocity vector with respect to Eurasia. We can easily observe the motion of convergence and the sudden decrease of north-eastward velocities moving from India to Tibet. Such a velocity field can be translated by spatial interpolation and derivation into a map of surface strain rate, as shown on the map on the right, highlighting the Himalayan region. GNSS data are used to derive a model of locking of the Main Himalayan Thrust. For a color version of this figure, see www.iste.co.uk/cattin/himalaya3.zip*

In contrast to earlier studies, we use a probabilistic approach to find the distribution of locking. Effectively, this problem can be solved via an optimization approach that will minimize the misfit between the model prediction (i.e. surface displacement rates that the model predicts) and the data (i.e. GNSS and leveling rates). Such methods are fast and robust but require some assumptions to regularize the search for the best solution. A model that would fit the data exactly despite the associated uncertainties would probably exhibit an almost random distribution of coupling values. Therefore, most optimization methods rely on smoothing of the solution to derive a final model that fits the data and is physically sound. A second type of method aims at exploring all the models that fit the data, considering the uncertainties, by sampling the posterior Probability Density Function (PDF) of the models given a data set. Often termed Bayesian, these methods do not require assumptions about the smoothness of the model and allow us to explore the uncertainties on the model. Prior assumptions are enclosed in the a priori PDF that describes our state of knowledge before considering any data. Sampling the posterior PDF allows us to evaluate how the data enables us to transition from this a priori state of information to the a posteriori data-driven state of information.

The latest published coupling model of the MHT is based on such approach (Figure 6.3; Dal Zilio et al. 2020) and consists of 360,000 models corresponding to the posterior information on geodetic coupling given measured interseismic velocities. Our a priori PDF assumes all models are equiprobable. Our final solution consists of an ensemble of models that are statistically distributed according to the posterior PDF without any assumption about the spatial distribution of locking for the given fault geometry.

6.2.3. *Interseismic coupling distribution*

Relaxing the smoothing assumption leads to an interesting picture. The posterior mean coupling model shows remarkable spatial variations in interseismic coupling, both down-dip and along-strike on the MHT (Figure 6.3). We infer discrete highly coupled segments along the fault extending throughout the arc. Coupling is high at the front, where the MHT reaches the surface and the resolution resulting from observations is highest. Down-dip, coupling tapers to negligible values around 100–120 km from the MHT. From west to east, we observe a large lateral variability in geodetic coupling. Four large (ca. 300–450 km) and discrete highly coupled segments

(>0.8) are separated by three relatively smaller (ca. 100 km) regions with lower coupling (<0.3).

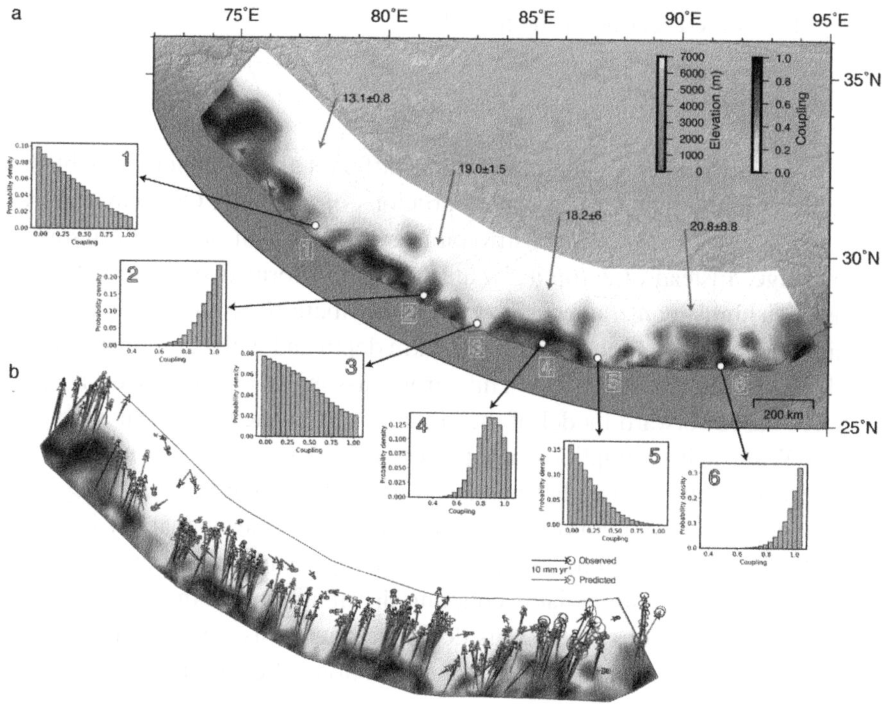

Figure 6.3. *(a) Distribution of interseismic coupling along the Main Himalayan Thrust and associated uncertainties. Color indicates coupling (i.e. the degree of locking of the megathrust). Red is fully locked while white means the MHT slips at the rate imposed by plate convergence. Histograms show the spread of values for various nodes of the fault mesh, highlighting the variable degree of confidence in the results. For instance, while node 5 is unlocked with relatively good confidence, node 3 is more uncertain although the mean of the models is low. (b) The blue arrows are the data and the black arrows are predictions from the mean model showing the good agreement between this solution and the GNSS data. For a color version of this figure, see www.iste.co.uk/cattin/himalaya3.zip*

The posterior PDF of coupling makes it possible to assess the robustness of our estimate by considering the local marginals of coupling (nodes 1–6 in Figure 6.3). Marginal PDFs for nodes 2, 4 and 6 show a >75% probability that fault coupling exceeds 0.8. However, for nodes 1 and 5, we infer a 68% probability that fault coupling is lower than 0.4, confirming a high probability of low coupling. In contrast, the shape of the cumulative distribution function of coupling on node 3 suggests a wider range of probable values, not far from the initial prior distributions (i.e. uniform probability of coupling) but still tending towards lower values of coupling. This similarity of prior and posterior distributions suggests the data are less informative for deriving fault coupling along this limited section of the model.

6.2.4. *Discussion*

This distribution of coupling is quite different from previously published models as it highlights along-strike variations of coupling. One way to assess the robustness of this latest distribution is to play with our a priori state of knowledge. Instead of assuming a uniform non-informative a priori PDF, we consider a highly constrained a priori PDF distribution, assuming that the MHT is highly coupled from the surface to 18 km depth and with negligible coupling at depths greater than 20 km. This strong assumption is based on a recent physics-based forward model of the Himalayan seismic cycle (Dal Zilio et al. 2019). With such assumption, the a posteriori PDF shows a mean model that is not very different from earlier published models assuming a certain degree of smoothness of coupling along the fault. Such result means that, by imposing smoothing on the solution, the optimization approach used earlier was forced to exclude models with variable coupling distribution from the list of possible models while a lot of such models do fit the data. This simple test highlights that the data prescribe such along-strike variations of coupling.

The spatial variability of interseismic coupling is an important factor to consider when interpreting relationships with the slip distribution of historical ruptures (Figure 6.4). Along-strike, low coupling regions around nodes 1, 3 and 5 coincide with three subsurface Indian basement ridges in the Indo-Gangetic plains: the Delhi–Haridwar ridge (DHR), the Faizabad ridge (FR) and the Munger–Saharsa ridge (MSR). These subsurface basement ridges extend northward across the Himalayan foothills, and they might have affected the development of the arcuate shape and the morphology of the Himalayan arc. They were detected from basin thickness changes (e.g. Gahalaut and Kundu 2012), and confirmed by the analysis of anomalous transient geomagnetic variations and arc-parallel gravity anomalies (Arora et al. 2012; Hetényi et al. 2016). They typically include oblique to transverse synclines, folds, fracture zones or faults transverse to the strike of the Himalaya, and they are associated with normal faulting and strike-slip seismicity (Valdiya 1976). The segmentation in the seismic behaviour of the MHT is likely related to the subduction of these major tectonic structures.

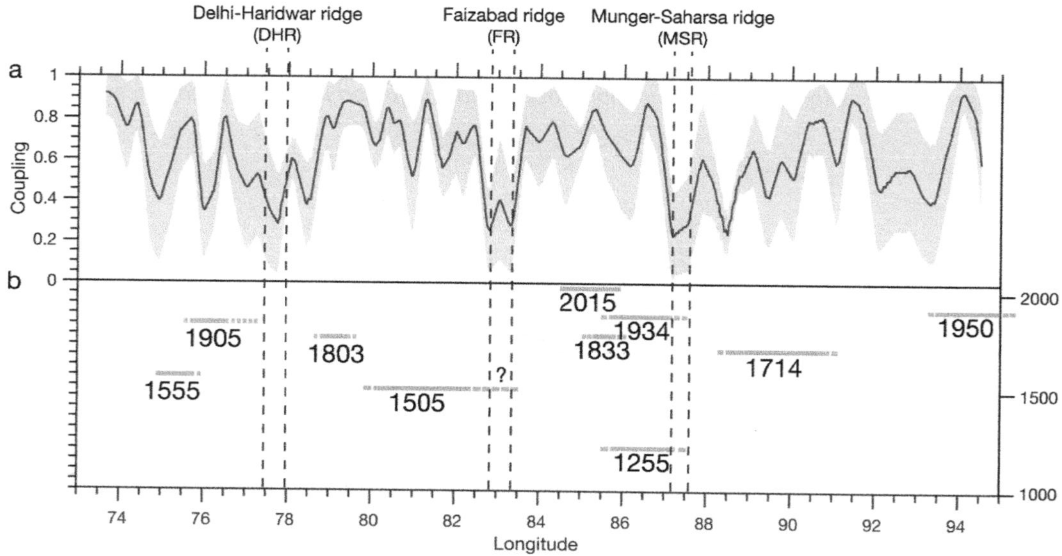

Figure 6.4. *Relationship between the along-strike variation of coupling (top) and the distribution of known seismic ruptures along the megathrust (bottom). The extent of ruptures is inferred from historical documents and paleo-seismological investigations (modified from Hetényi et al. 2016). Dashed blue lines indicate the position of structural highs in the subducting Indian basement: the Delhi–Haridwar, the Faizabad and the Munger–Saharsa ridges. For a color version of this figure, see www.iste.co.uk/cattin/himalaya3.zip*

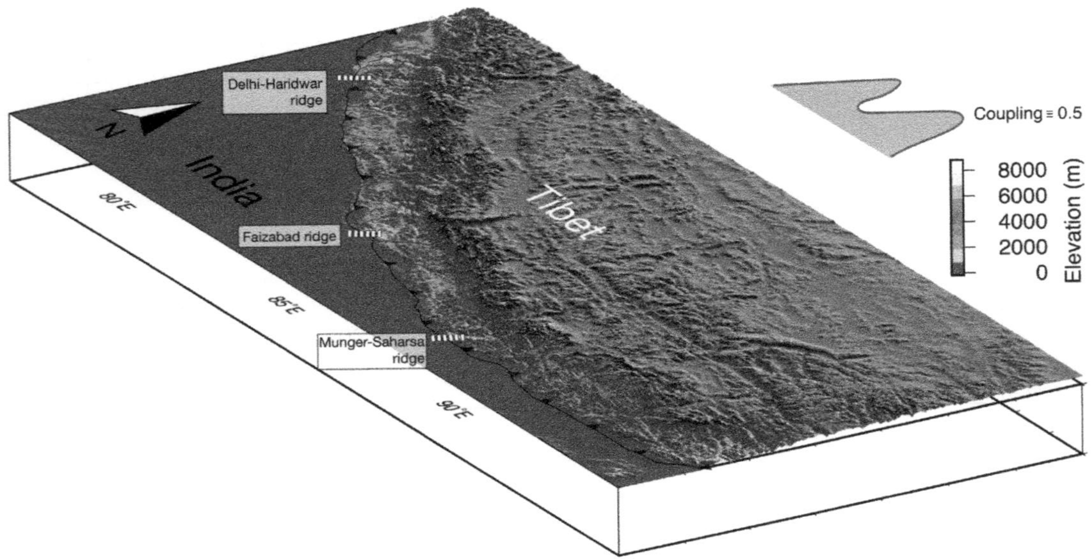

Figure 6.5. *Spatial extent of potential rupture sizes along the entire Himalayan arc (pink shading). These areas can be used as potential sources for the definition of seismic hazard in the region. For a color version of this figure, see www.iste.co.uk/cattin/himalaya3.zip*

To conclude on the distribution of coupling along the Main Himalayan Thrust, we could also discuss the potential relationship with the building of topography and the distribution of the gravity anomaly (Dal Zilio et al. 2020), but this would take us away from the evaluation of the seismic potential in the Himalaya. Coupling along the MHT is variable and defines regions of increased hazard (Figure 6.5). These regions have experienced large earthquakes in the past and the relationship between the extent of past ruptures and coupling is quite clear. However, how can we translate such evaluation of the degree of locking along the megathrust into meaningful information for the assessment of seismic hazard?

6.3. Seismic potential of the MHT

GNSS data suggest that the MHT is to the first order locked from 15 km depth to the surface (Stevens and Avouac 2015) but shows heterogeneous coupling along-strike (Dal Zilio et al. 2020). In particular, three regions with potentially low coupling, located at structural geological features, might act as frictional barriers to the propagation of earthquakes. In this section, we attempt to provide an estimate of the seismic potential of the MHT. In particular, we will address the question of the potential maximum magnitude earthquake, M_{max}, and its associated recurrence time, τ_{max}, both of which must be affected by along-strike heterogeneities.

The seismic potential of a fault can be evaluated in a probabilistic fashion based on earthquake statistics and additional physical constraints. In the following, we consider four constraints: (1) the moment conservation principle; (2) the seismicity magnitude–frequency distribution; (3) the earthquakes' magnitude-rupture area scaling law; and (4) the effect of along-strike frictional barriers based on statistical results from simulations of earthquake cycles. Each of these constraints described in the following sections are characterized in Michel et al. (2021), and are also based on previous work on seismic potential of different regions of the world (e.g. Stevens and Avouac 2016; Michel et al. 2018; Rollins and Avouac 2019).

6.3.1. *Conservation of the seismic moment*

During the interseismic period, the sections of fault that are locked accumulate elastic strain, as described previously. The elastic strain accumulated through time is then released in various ways, among them earthquakes (Reid 1910). Other outcomes include elastic strain release through aseismic slip or plastic deformation. Assuming the crust is to first-order elastic, all strain measured in the interseismic period must be balanced by slip on faults, either coseismic (i.e. during the earthquake) or post-seismic (i.e. after the earthquake). It is possible to evaluate and quantify this equilibrium by evaluating the budget of seismic moment, m_0, on a given fault (Avouac 2015b).

On the one hand, the seismic moment of an earthquake is defined as the integral of slip over a given fault area, such as

$$m_0^{seis} = \int \mu D^{seis} dA, \tag{6.1}$$

where μ is the shear modulus, A is the area over which the fault slipped and D^{seis} is the seismic slip.

In addition, we know that aseismic afterslip generally occurs following an earthquake, and that earthquakes are followed by sequences of aftershocks, each of them releasing a fraction of the available moment. In addition to large events and their post-seismic sequences, background seismicity is assumed to be the sum of independent events and their recurrence time is Poissonian. Each of the events in background seismicity is also followed by some post-seismic moment release. Over a given period of time, the total moment release rate is then the sum of all coseismic and post-seismic moment release. For large events, we can measure the post-seismic moment released by large events with geodetic and seismological techniques, and it typically amounts to 25% of the coseismic moment release. For smaller events, the ratio between the post-seismic and co-seismic moment release typically ranges from 0% to more than 100% (Avouac 2015b; Alwahedi and Hawthorne 2019).

On the other hand, during the interseismic period, moment deficit builds up at a rate, $\dot{m}_0^{def}$, defined as the integral of the slip deficit rate over the fault area,

$$\dot{m}_0^{def} = \int \mu \dot{D}^{load} dA, \tag{6.2}$$

where $\dot{D}^{load}$ is the slip deficit rate, the rate at which a locked asperity is loaded. $\dot{D}^{load}$ is directly related to local coupling as

$$\dot{D}^{load} = \chi V_{plate}, \tag{6.3}$$

where χ is the interseismic coupling and V_{plate} is the long-term plate rate. Over the long-term, if we assume the moment released through plastic deformation and aseismic transients during the interseismic period is negligible, the moment release rate, $\dot{m}_0^{Total}$, should be equal to $\dot{m}_0^{def}$. Building a budget of moment hence allows us to evaluate how much slip over a given area there is to be released, which is directly related to the potential magnitude of future earthquakes.

Evaluating the moment deficit rate of the MHT can be done using the coupling map described in the previous section (Dal Zilio et al. 2020). The 360,000 generated coupling models sampled from the posterior PDF of coupling each correspond to a given moment deficit rate, hence we obtain the PDF of moment deficit buildup rate. Moment deficit buildup rate is normally distributed with a mean and standard deviation of $9.88 \pm 0.29\ 10^{19}$ N.m.yr^{-1}. Such a moment buildup rate corresponds to a seismic deficit of magnitude 7.2 every year (defined as moment magnitude, Hanks and Kanamori 1979).

6.3.2. *Magnitude–frequency distribution in the Himalaya*

It is important to realize that the instrumental observation period of earthquakes ($\sim$100 years) is short compared to the recurrence time of large earthquakes (M>8) which only occur every few hundred years or so. The exact total moment budget is therefore difficult to estimate, if not impossible. Therefore, we instead build seismicity models representing the earthquake productivity of a given region. Such productivity can be described by a magnitude–frequency distribution (MFD) of background seismicity. It has now been empirically demonstrated that earthquakes follow a form of power-law distribution that roughly states there are 10 times more earthquakes of magnitude M than earthquakes of magnitude M+1. Empirically, the cumulative distribution of earthquakes as a function of magnitude shows a linear trend with a slope b controlling the relative proportion of earthquakes of each magnitude. This law is often referred to as the Gutenberg–Richter law. We can easily see that such a distribution will allow us to evaluate the proportion of earthquakes of each magnitude that will occur in the future for a given total moment to be released. It is therefore important to capture the shape of the magnitude–frequency distribution for the Himalaya.

We assume that the MFD follows a power-law form which tapers to a certain maximum magnitude, M_{max}. Additional forms can be tested and their effect is evaluated in Michel et al. (2021). Our seismicity model is thus described by a set of four parameters: (1) the maximum magnitude earthquake, M_{max}, that can occur on a fault; (2) the recurrence rate of events of a certain magnitude, τ_c, that will then determine the rates of the whole seismicity model; (3) the relative rate between small and large events, b (the slope of the Gutenberg–Richter law) and (4) the proportion of moment released post-seismically (aftershocks and aseismic afterslip) relative to the total moment released, α.

The moment release rate of each model, $\dot{m}_0^{Total}$, can then be evaluated and compared to the moment deficit build-up rate, $\dot{m}_0^{def}$. In particular, the cumulative MFD that balances the moment budget is given as follows:

$$N(> M_w) = \frac{1 - \frac{2b}{3}}{\frac{2b}{3}} \times \frac{\alpha_s \dot{m}_0^{def}}{m_0^{max}} \times \left[\left(\frac{m_0^{max}}{m_0(M_w)} \right)^{\frac{2b}{3}} - 1 \right], \qquad [6.4]$$

where $N(> M_w)$ is the rate of $\geq M_w$ events, m_0^{max} is the moment corresponding to the maximum magnitude earthquake, M_{max}, and $m_0(M_w)$ is the moment of any earthquake of magnitude M_w (Rollins and Avouac 2019).

Therefore, as we have the PDF of the moment deficit rate from the coupling model, we only need the probability of a given seismic model that depends on the PDFs of M_{max}, the maximum magnitude in the region, b, the slope of the Gutenberg–Richter law in the region and α_s, the proportion of post-seismic moment release.

In order to constrain these parameters, we compare these seismicity models with the catalogue of seismicity derived from seismic data in the Himalaya. Large events (Mw >7.7) are inferred mostly from historical records, paleo-seismological data and current instrumental data (Bilham 2019). Smaller events, from Mw 4.5 to 7.7, are from the ANSS catalogue. We compare these instrumental and historical catalogs, respectively, with the long-term seismicity MFD models. Effectively, we generate random catalogs of earthquakes assuming seismicity obeys a Poisson distribution and compare the observed MFD with the one randomly generated from our MFD. The probability of a given seismicity model then depends on the moment buildup rate (i.e. this MFD must close the moment budget) and on the various parameters of the MFD and in particular the maximum magnitude. More details can be found in Michel et al. (2021).

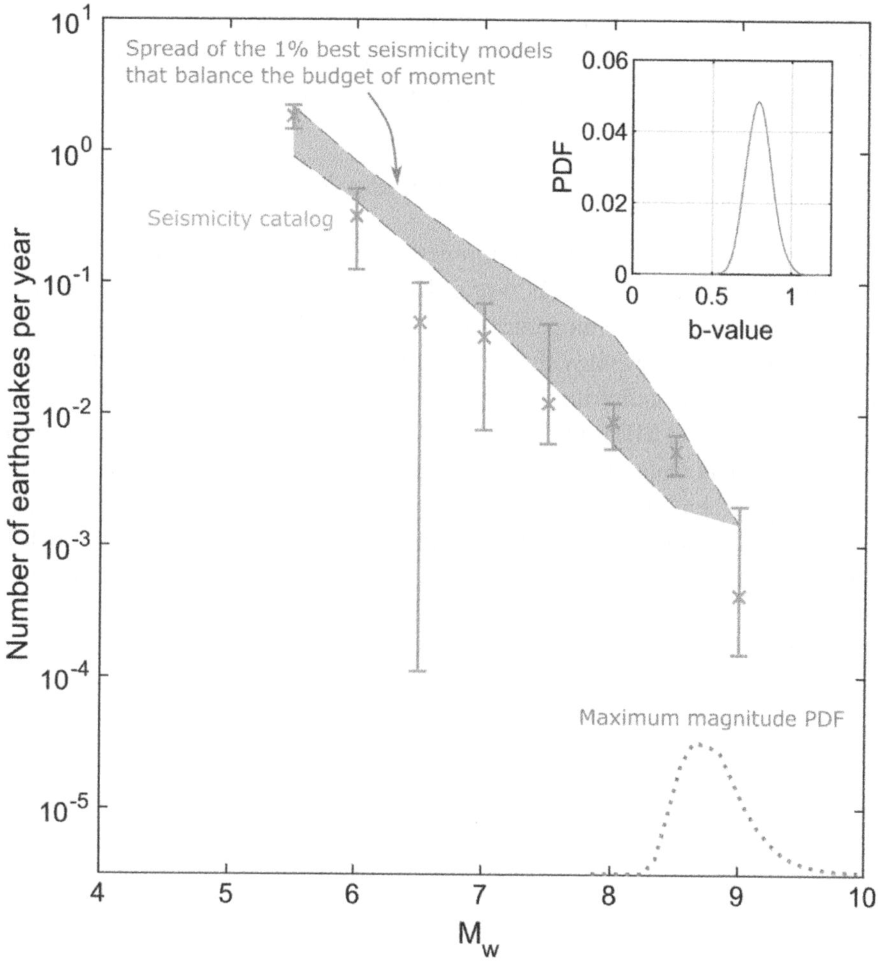

Figure 6.6. *Probability density functions resulting from the moment budget evaluation. The orange crosses and error bars correspond to the current and historical seismicity catalog and their associated uncertainties (Bilham 2019, ANSS). Green shading indicates the spread of the 1% best seismicity models that balance the budget of moment. The dotted green line on the x-axis is the marginal PDF of maximum magnitude that defines the seismicity models. Inset shows the marginal PDF of the b value of the same seismicity models. For a color version of this figure, see www.iste.co.uk/cattin/himalaya3.zip*

We therefore build a distribution of seismicity models that balance the budget of moment (Figure 6.6) and the corresponding distribution of b values for the region. In addition, this evaluation of the balance between moment buildup rate and observations of seismicity in the region suggest the probability of M_{max} starts to rise around Mw ~8.1, peaks at Mw ~8.7.

6.3.3. *Including the physics of fault slip in seismic hazard*

Global statistics show that the moment released by earthquakes scales with their rupture area: $m_0^{seis} \propto A^{3/2}$ (e.g. Kanamori and Brodsky 2004; Ye et al. 2016). However, the seismogenic zone of the MHT is defined by a finite area and can only host an event with a certain rupture area. Considering such scaling law, it is therefore quite unlikely that an event of magnitude larger than 9.5 should rupture the MHT in the future. We can thus test a set of events from Mw 6 to 10, calculate their moment released, m_0^{seis}, and evaluate the area they would rupture based on the moment-area scaling law. If an event does not fit within the seismogenic zone, then such event is improbable. For the MHT, taking a coupling threshold of 0.3 as the limit of the seismogenic zone, such a constraint affects the probability of events with magnitude above Mw 9.1 (Michel et al. 2021). We will include this consideration in the probability of occurrence of the different magnitude earthquakes.

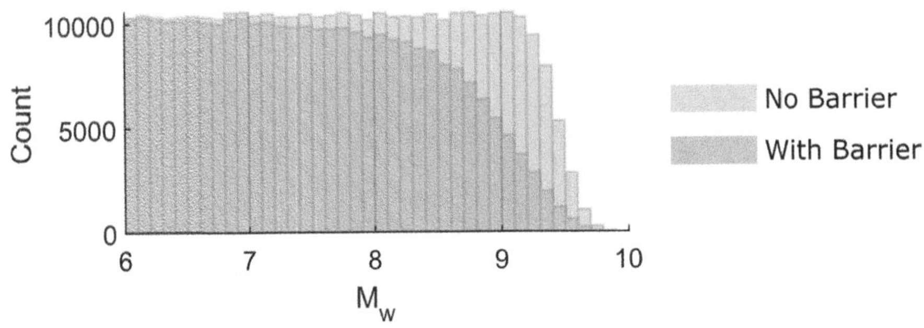

Figure 6.7. *Effect of empirical earthquake scaling law and frictional barriers on the probability of earthquake occurrence. Pink histograms shows the probability of number of earthquakes per magnitude bin that passed the scaling law test. Yellow histogram shows the number of earthquakes per magnitude bin that passed the scaling law test and the frictional barrier test. We observe that earthquakes of magnitude smaller than 7 are not affected by these constraints while events of magnitude larger than 9.6 are implausible. For a color version of this figure, see www.iste.co.uk/cattin/himalaya3.zip*

In addition to such empirical scaling, it has been proposed that frictional heterogeneities might affect the propagation, hence the final size, of earthquakes. The distribution of coupling along the MHT shows three barriers, which can be interpreted as frictional barriers, separating locked asperities where earthquakes might happen. To evaluate the effect of such barriers, we use results based upon dynamical simulations of the seismic cycle from Kaneko et al. (2010). This study defined the probability of a seismic event to pass a frictional barrier as dependent on a non-dimensional parameter called the barrier efficiency, B. This parameter, which depends on the frictional and stress conditions acting on the fault, has been investigated using numerical simulations relating the barrier efficiency B with the probability that an earthquake will pass a barrier (i.e. B close to 0 corresponds to events crossing easily through the barrier while $B > 1$ tends to block any seismic event). For the Himalayan case, all the parameters constitutive of B, and their related uncertainties, are either available from the Dal Zilio et al. (2020) coupling model, through other studies on the MHT or from global earthquake statistics. We can then test whether an event of a certain magnitude, randomly fitted within the seismogenic zone, and with parameters of B sampled within their uncertainties, can propagate through the MHT potential barriers (Michel et al. 2021). We therefore generate a large number of earthquakes randomly, check whether they follow the scaling law and check whether they cross the barriers we have identified or not (Figure 6.7).

The results show that Mw $>$8 events begin to pass one barrier and thus have a harder time to propagate (Michel et al. 2021). The larger an event, the more barriers it needs to pass, and the less probable they become considering this framework. However, the barriers do not manage to drop the probability of a full-length rupture of the MHT to 0, since around a quarter of such tested events can still break through all barriers.

6.3.4. *Seismic potential of the MHT*

Performing such exploration allows us to derive several numbers that will be useful for probabilistic seismic hazard assessments. First, we obtain a more constrained PDF of the maximum magnitude to be expected in the region (Figure 6.8). The maximum magnitude is 8.6 with a smaller spread and the recurrence time of any events with similar magnitude, aftershocks excluded, is estimated at 1,000 years.

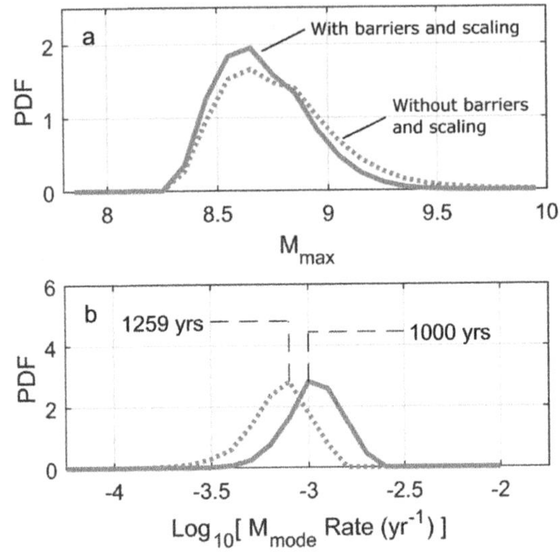

Figure 6.8. *(a) Comparison of the PDF of the maximum magnitude that can rupture the MHT accounting for (continuous lines) and without (dashed lines) the effect of frictional barriers and scaling laws. Including barriers and scaling laws reduces slightly the probability of the maximum magnitude earthquake for magnitudes larger than 8.8. Consequently, the recurrence time of background events with magnitude below 8.8 become statistically more frequent. This is illustrated in panel (b) which shows the inferred rate of Mw 8.6 events. For a color version of this figure, see www.iste.co.uk/cattin/himalaya3.zip*

Among the limitations of the current approach, it is important to note that we assume a Poissonian distribution for earthquake productivity as we consider declustered catalogues (i.e. aftershocks have been removed). However, in doing so, the methodology does not account for the timing of events and is thus time independent. This is a major assumption as large earthquakes will strongly affect the state of stress along the MHT, which has been shown to impact the timing and the size of subsequent events.

In addition, we consider the barriers to be frictional ones, but the geometry of the MHT also has an influence on the segmentation of the MHT. For instance, the 2015 Mw 7.8 Gorkha earthquake ruptured partially the locked zone of the MHT and was constrained to the lower edge of the seismogenic zone (Avouac et al. 2015; Galetzka et al. 2015; Elliott et al. 2016) potentially due to the stress shadow left by previous earthquakes (Michel et al. 2017; Dal Zilio et al. 2019) or due to an intermediate geometrical ramp on the MHT

(Hubbard et al. 2016). Such considerations on the physics of fault slip in relationship to geological and historical elements are missing and should be considered for future earthquake hazard assessments.

With the results and assumptions of this section in mind, we can now move to the evaluation of seismic hazard using ground motion prediction equations to derive maps of peak ground acceleration expected at certain probability levels during the next decades.

6.4. Seismic hazard in the Himalaya

Now that the parameters of the magnitude–frequency distribution and the maximum magnitude earthquake on the MHT have been constrained, we need to define other faults or earthquake source areas and combine all the earthquake sources with Ground Motion Prediction Equations (GMPEs) to calculate seismic hazard in a probabilistic way. The results of Probabilistic Seismic Hazard Assessment (PSHA) are often given in terms of the level of shaking that would be reached with different probabilities in a certain time period. For example, the results might show the level of shaking that might be reached with a 10% chance in a 50-year period. Given their nature, there is a higher probability of smaller magnitude earthquakes, which means that in a 50-year period there might be a large chance that a low level of shaking will be reached. There will be a much smaller chance that very high levels of shaking will be reached. PSHA tries to quantify this.

In addition to PSHA, scenario-based hazard analysis can be performed. This involves modeling a specific earthquake rupture or scenario (often based on past large events) and estimating the shaking produced. This is useful for several reasons: the results of scenario models are often easier to understand (especially for policy-makers) since they do not involve probabilities, they can give an idea of the size of a region that might be affected by a single event, and if based on historical events, the estimated shaking can be compared with real data from that event. However, there are also several drawbacks associated with rupture scenarios: no probabilities are associated with each scenario and the next large earthquake in the region may be very different to any of the scenarios studied, and GMPEs have large uncertainties which are sampled so the shaking results sampled once may not correspond to the real observations.

6.4.1. *Ground motion prediction equations and Vs30*

Ground Motion Prediction Equations (GMPEs) are designed to estimate the amount of shaking at different locations as a result of an earthquake. The nearer a location is to an earthquake rupture, and the larger the amount of slip on the fault during the rupture, the greater the shaking will be. GMPEs have the basic form:

$$\log(PGA) = a_1 + a_2 \times M - a_3 \times R \quad [6.5]$$

where PGA stands for the peak ground acceleration (often given in terms of gravity, g), M is the magnitude of the considered earthquake, R is the distance to the rupture, and a_1, a_2 and a_3 are the empirically found constants.

Figure 6.9 shows an example of the results from such an equation. In addition, site characteristics such as the properties of the soil or bedrock at each location are included in many GMPEs in the form of the Vs30 value for each site. This is the average shear-wave velocity from the ground surface to 30 m depth and can be found from geophysical methods such as boreholes. Proxy Vs30 can also be estimated from topographic slope (Wald and Allen 2007; Allen and Wald 2009). Steeper slopes are more likely to be solid bedrock (higher Vs30) than flat areas (lower Vs30) which are likely to be covered with sediments.

GMPEs have been developed both for regional and global use, and generally are designed for specific tectonic regions such as subduction zones, or active continental crust. In the Himalayan region, there have been a few local studies, so GMPEs developed from global datasets have to be used. However, based on shaking data from the 2015 Gorkha earthquake, GMPEs have now been developed for Kathmandu Valley (Mori et al. 2020). In the following, we use the GMPEs of Zhao et al. (2006), Boore et al. (2014) and Abrahamson et al. (2016). These were chosen as they are globally applicable and designed for subduction zones and active continental crust.

There are also few local Vs30 measurements in the Himalayan Region, although there have been some local studies for Kathmandu Basin and some cities in northern India (Nhemafuki 2008; Pandey et al. 2016; Tallett-Williams et al. 2016; Pudi et al. 2021). These local measurements compare favorably with Vs30 values from the USGS (https://earthquake.usgs.gov/data/Vs30/) which are used in the subsequent analysis. In the future as more local studies are undertaken, the results of these could be compared with different methods of estimating proxy Vs30 values (from geology, terrain or topographic slope) to find the best estimator of Vs30 values (e.g. Wu et al. 2020).

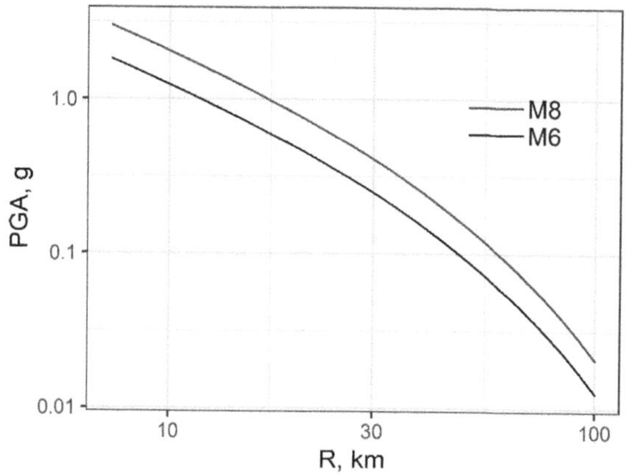

Figure 6.9. *Example PGA (Peak Ground Acceleration) versus distance to a rupture R for an Mw 6 and an Mw 8 earthquake. Note that both axes are in logarithmic scale. For a color version of this figure, see www.iste.co.uk/cattin/himalaya3.zip*

Cities in the region are typically located in valleys and basins, where Vs30 is low, leading to amplification of shaking. Dehradun in northern India and Kathmandu in Nepal are both large cities in basins with low Vs30 values and Thimphu, a smaller city and the capital of Bhutan, is located in a valley between the mountains which also has low Vs30 values (Figure 6.10). GMPEs are constantly being improved and designed to take into account of more and more parameters such as fault type, directivity and geometries of sedimentary basins.

6.4.2. *Modeling scenario events*

The most recent large event in the Himalayan region is the 2015 Mw7.8 Gorkha Nepal earthquake (see Volume 3 – Chapter 8). We can use the rupture geometry and extent together with the magnitude of this event to model expected ground shaking and compare with real observations. This also involves choosing GMPEs. As noted in the previous section, no regional GMPEs have been developed for Nepal so we choose several global GMPEs and use each with equal probability. Although we observe quite some variation between the three GMPEs tested here, the average values compare reasonably well with observations (Figure 6.11).

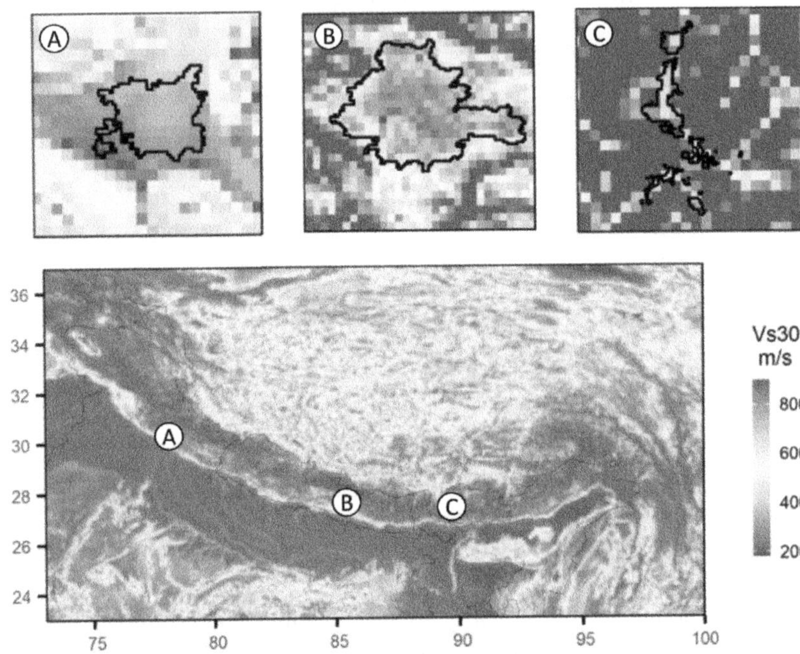

Figure 6.10. *Proxy Vs30 values from the USGS and details from left to right for Dehradun, India (A), Kathmandu, Nepal (B) and Thimphu, Bhutan (C). For a color version of this figure, see www.iste.co.uk/cattin/himalaya3.zip*

In the Kathmandu basin itself, much lower values of shaking are observed than expected. For example, the soft-sediment site at KATNP recorded a PGA of 0.16 g (Bhattarai et al. 2015), while the models would suggest around 0.55 g, and at the hard-rock sediment site of KTP a PGA value of 0.25 g was observed Takai et al. (2016), while the models also suggest around 0.5–0.55 g. This is partly because for this specific earthquake, the areas of high slip on the fault and in particular sources of high-frequency shaking were along the northern edge of the fault rupture, further from Kathmandu (Avouac et al. 2015). The lower shaking in Kathmandu may also have been partly due to the geometric shape of the basin and the characteristics of the sediments within it (Asimaki et al. 2017). In our simple scenario model, slip is equally likely to occur anywhere on the rupture and is expected to cause greater shaking above the southern end of the rupture because this is at a shallower depth leading to lower attenuation.

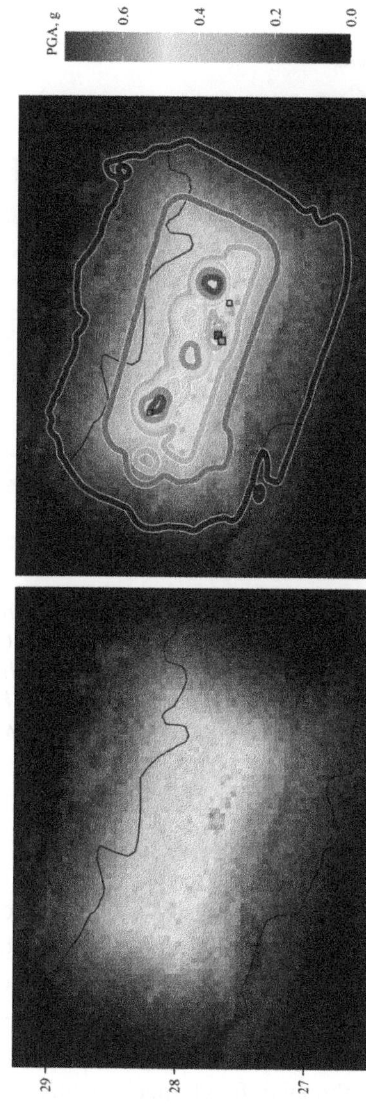

Figure 6.11. *Average predicted Peak Ground Acceleration from a single scenario using three different GMPEs (left). Colored lines (right) show the predicted PGA map produced by USGS. Colored squares show PGA values measured at two seismometers in Kathmandu basin in 2015. For a color version of this figure, see www.iste.co.uk/cattin/himalaya3.zip*

While modeling the Gorkha earthquake is useful to compare the results, it does not by itself tell us anything about the probability of future shaking. For that we also have to consider the probabilities of earthquakes and their potential locations, as we study in the following section.

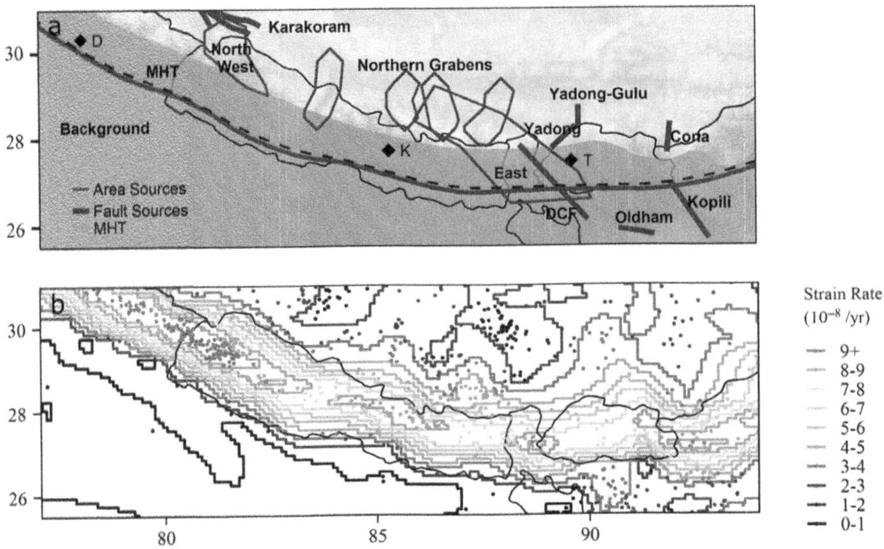

Figure 6.12. *Seismic source areas for the central Himalayan region based on (a) faults and areas of similar tectonics and (b) strain rate from GNSS. Earthquakes are color-coded by the area of strain rate that they fall within. D = Dehradun, K = Kathmandu, T = Thimphu, DCF = Dhubri-Chungthang Fault, MHT = Main Himalayan Thrust. For a color version of this figure, see www.iste.co.uk/cattin/himalaya3.zip*

6.4.3. *Probabilistic seismic hazard assessment*

For probabilistic seismic hazard analysis, we now need to consider the probabilities that earthquakes occur in different regions, and what magnitudes these earthquakes will have. We have shown how we can evaluate such quantities using geophysical observations. Here, we use source regions for the central Himalaya from Stevens et al. (2018, 2020), who defined source regions on the basis of earthquake focal mechanisms and the location of faults seen in the landscape and through seismic studies (Figure 6.12). We also test an alternative method used by Stevens and Avouac (2021) who used strain rate to automatically define regions rather than a subjective classification (Figure 6.12).

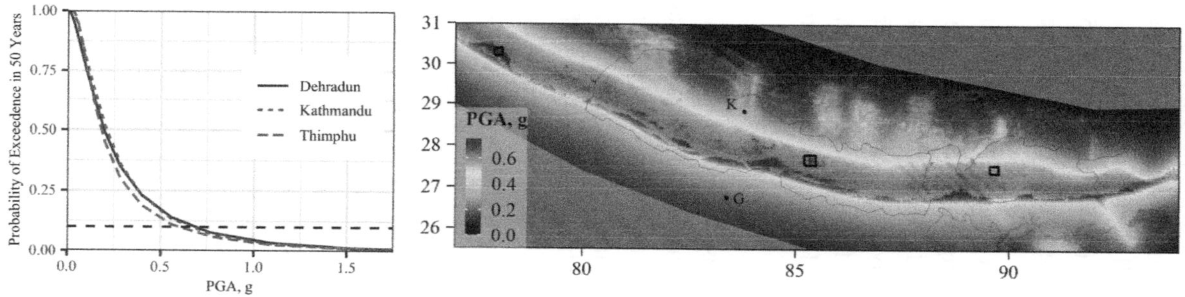

Figure 6.13. *Right: PSHA map showing the 10% chance that a given level of shaking (i.e. PGA) is reached in 50 years. Left: Exceedance curves for the cities of Dehradun, Kathmandu and Thimphu (shown as dark squares on left map). Dashed lines shows the 10% probability of exceedance line which is shown for all locations on the map. For a color version of this figure, see www.iste.co.uk/cattin/himalaya3.zip*

The probability that certain locations will reach a certain level of shaking is a combination of the probability of earthquakes of different sizes occurring and the probability that each such earthquakes will cause a certain level shaking at the location of interest. There are a few open-source tools available to perform the PSHA calculation once we have all the input data needed. For example, OpenQuake (Pagani et al. 2014), CRISIS (Ordaz et al. 2015) and OpenSHA (Field et al. 2003) are all freely available to use. Here, we use OpenQuake.

Figure 6.13 shows the results of a PSHA analysis using the manually chosen area and fault sources while Figure 6.14 shows the results of using the strain rate-based areas. As expected, the Main Himalayan Thrust, which underlies most of the Central Himalayan region, is the greatest source of hazard. Grabens on the southern edge of the Tibetan Plateau, and some faults that extend south of the Himalaya in Bhutan are also sources of increased hazard.

The results from the strain area-based method show a lower hazard in this region although with a similar overall shape. This is partly because the geometry of the MHT is not an input to this model and so earthquakes are not constrained to rupture along this fault plane. In this model, they have an equal chance of rupturing in any direction, which is a limitation of the model since we know this is not physically possible. The consequences are that the area of higher hazard is spread over a larger region instead of concentrated above the fault plane itself.

There are advantages and disadvantages to both of these methods. The strain-based method has less subjective human judgment and can cover a large area easily in a uniform manner, and it is also easy to update as new GNSS and earthquake data become available. However, as mentioned above, this method does not make use of some knowledge that we do have such as the geometry and slip rates on faults. The large differences between these two models show the large uncertainties related to PSHA, not only from uncertainties of the parameters within each model, but also from uncertainties to do with the modeling techniques themselves.

Time dependency is not considered here. If there has recently been a very large earthquake on a fault, then it might be less likely that another such large earthquake will occur in the same location in the near future. On the contrary, if some regions have not had a large earthquake in a long time, it might be that a large earthquake is more likely to happen here. For the Himalayan Region, the 2015 Mw 7.8 Gorkha earthquake, while large, was not at all as large as earthquakes that have occurred on the MHT this region in the past, for example, the 1934 or 1950 events, and paleoseismic events that may be around M9.

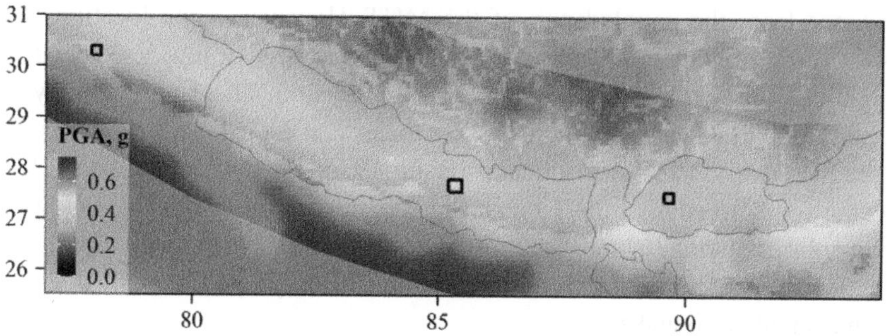

Figure 6.14. *PSHA results showing the 10% chance that these levels of shaking are reached in 50 years using the method based on strain rate areas rather than fault and area sources. The cities of Dehradun, Kathmandu and Thimphu are shown as dark squares. For a color version of this figure, see www.iste.co.uk/cattin/himalaya3.zip*

6.4.4. *From hazard to risk*

So far we have considered hazard. Risk is a combination of hazard, exposure and vulnerability. Exposure includes the number of people and buildings that might be affected by the hazard, and vulnerability is how strongly these people and buildings are affected. Hazard and risk do not necessarily correlate, as a high hazard in the middle of an uninhabited area would lead to a low risk, while a medium-sized hazard in a heavily populated urban area would lead to a high risk. While hazard cannot be changed, as in we cannot reduce the number of earthquakes, risk can be mitigated by not building in hazardous areas and retrofitting buildings in hazardous areas so that they are less likely to collapse or become damaged in earthquakes.

While hazard levels in the Himalaya remain fairly constant, risk is increasing because of population growth, urbanization, increasing inequality, and economic growth leading to more infrastructure within the region. Identifying the seismic hazard as discussed in this section is only the first step in beginning to mitigate the risk in this region.

6.5. Conclusion

Hazard is high in the Himalaya although it does vary spatially. The MHT is the main source of seismic hazard and hazard generally decreases with distance from the coupled part of the MHT. However, some locations such as those on the Ganges Plain have thick sediments, so a low Vs30 value, so might be particularly affected by local amplification. These regions may also experience liquefaction, which has not been considered here, but was seen after the Mw 8.2 1934 earthquake (Rana 1935; Pandey and Molnar 1988). On the southern edge of the Tibetan Plateau, the earthquake hazard is lower in many places, although landslides might be a cause of concern, and limited road access means these remote communities may be disproportionally affected by even a small earthquake.

To further constrain the hazard in the future, denser networks of GNSS and complementary InSAR studies would help to refine the rate of seismic moment buildup in the area, and also help to constrain the slip rate on faults smaller than the MHT. A network of broadband seismometers that record shaking from earthquakes is particularly important to install as soon as possible to create GMPEs specific to the region. Dense geophysical Vs30 studies in heavily populated regions such as the Kathmandu valley might allow for seismic microzonation, that is, avoid building on locations that are particularly susceptible to high shaking during earthquakes. In addition, improving the evaluation of the probability of each scenario will require better physical characterization of the structures involved and physics-based models of the earthquake cycle are key here. We will conclude this chapter by stating that, although risk can be mitigated to some level, seismic hazard and all subsequent related hazards in the Himalaya is high and will remain as such at all times.

6.6. References

Abrahamson, N., Gregor, N., Addo, K. (2016). BC hydro ground motion prediction equations for subduction earthquakes. *Earthquake Spectra*, 32(1), 23–44.

Ader, T., Avouac, J., Liu-Zeng, J., Lyon-Caen, H., Bollinger, L., Galetzka, J., Genrich, J., Thomas, M., Chanard, K., Sapkota, S.N. et al. (2012). Convergence rate across the Nepal Himalaya and interseismic coupling on the Main Himalayan Thrust: Implications for seismic hazard. *Journal of Geophysical Research*, 117(B04403), 1–16.

Allen, T.I. and Wald, D.J. (2009). On the use of high-resolution topographic data as a proxy for seismic site conditions (Vs30). *Bulletin of the Seismological Society of America*, 99(2A), 935–943.

Alwahedi, M.A. and Hawthorne, J.C. (2019). Intermediate-magnitude postseismic slip follows intermediate-magnitude (M 4 to 5) earthquakes in California. *Geophysical Research Letters*, 46(7), 3676–3687.

Armijo, R., Tapponnier, P., Mercier, J.L., Han, T.-L. (1986). Quaternary extension in southern Tibet: Field observations and tectonic implications. *Journal of Geophysical Research*, 91(B14), 13803.

Arora, B.R., Gahalaut, V.K., Kumar, N. (2012). Structural control on along-strike variation in the seismicity of the northwest Himalaya. *Journal of Asian Earth Sciences*, 57, 15–24.

Asimaki, D., Mohammadi, K., Mason, H.B., Adams, R.K., Rajaure, S., Khadka, D. (2017). Observations and simulations of basin effects in the Kathmandu Valley during the 2015 Gorkha, Nepal, earthquake sequence. *Earthquake Spectra*, 33(S1), S35–S53.

Avouac, J.-P. (2015). From geodetic imaging of seismic and aseismic fault slip to dynamic modeling of the seismic cycle. *Annual Review of Earth and Planetary Sciences*, 43(1), 233–271.

Avouac, J.-P., Meng, L., Wei, S., Wang, T., Ampuero, J. (2015). Lower edge of locked Main Himalayan Thrust unzipped by the 2015 Gorkha earthquake. *Nature Geoscience*, 8(9), 708–711.

Bhattarai, M., Adhikari, L.B., Gautam, U.P., Laurendeau, A., Labonne, C., Hoste-Colomer, R., Sèbe, O., Hernandez, B. (2015). Overview of the Large 25 April 2015 Gorkha, Nepal, earthquake from accelerometric perspectives. *Seismological Research Letters*, 86(6), 1540–1548.

Bilham, R. (2019). Himalayan earthquakes: A review of historical seismicity and early 21st century slip potential. *Geological Society, London, Special Publications*, 483(1), 423–482.

Bilham, R. and England, P. (2001). Plateau "pop-up" in the great 1897 Assam earthquake. *Nature*, 410(6830), 806–809.

Bilham, R., Mencin, D., Bendick, R., Bürgmann, R. (2017). Implications for elastic energy storage in the Himalaya from the Gorkha 2015 earthquake and other incomplete ruptures of the Main Himalayan Thrust. *Quaternary International*, 462, 3–21.

Bollinger, L., Sapkota, S. N., Tapponnier, P., Klinger, Y., Rizza, M., Van der Woerd, J., Tiwari, D.R., Pandey, R., Bitri, A., Bes de Berc, S. (2014). Estimating the return times of great Himalayan earthquakes in eastern Nepal: Evidence from the Patu and Bardibas strands of the Main Frontal Thrust. *Journal of Geophysical Research: Solid Earth*, 119(9), 7123–7163.

Boore, D.M., Stewart, J.P., Seyhan, E., Atkinson, G.M. (2014). NGA-West2 equations for predicting PGA, PGV, and 5% damped PSA for shallow crustal earthquakes. *Earthquake Spectra*, 30(3), 1057–1085.

Cattin, R. and Avouac, J. (2000). Modeling mountain building and the seismic cycle in the Himalaya of Nepal. *Journal of Geophysical Research*, 105(B6), 13,989–13,407.

Chen, W. and Molnar, P. (1990). Source parameters of earthquakes and intraplate deformation beneath the Shillong Plateau and the Northern Indoburman Ranges. *Journal of Geophysical Research*, 95(B8), 12527.

CIESIN (2009). Gridded population of the world version 3 (GPWv3): Population density grids. Socioeconomic Data and Applications Center (SEDAC), Columbia University, Palisades.

Copley, A., Avouac, J.-P., Wernicke, B.P. (2011). Evidence for mechanical coupling and strong Indian lower crust beneath southern Tibet. *Nature*, 472(7341), 79–81.

Dal Zilio, L., van Dinther, Y., Gerya, T., Avouac, J.-P. (2019). Bimodal seismicity in the Himalaya controlled by fault friction and geometry. *Nature Communications*, 10(1), 48.

Dal Zilio, L., Jolivet, R., Dinther, Y. (2020). Segmentation of the main Himalayan thrust illuminated by Bayesian inference of interseismic coupling. *Geophysical Research Letters*, 47(4), 1–10.

Dal Zilio, L., Hetényi, G., Hubbard, J., Bollinger, L. (2021). Building the Himalaya from tectonic to earthquake scales. *Nature Reviews Earth Environment*, 2(4), 251–268.

Elliott, J.R., Jolivet, R., González, P.J., Avouac, J., Hollingsworth, J., Searle, M.P. (2016). Himalayan megathrust geometry and relation to topography revealed by the Gorkha earthquake. *Nature Geoscience*, 9, 174–180.

Field, E.H., Jordan, T.H., Cornell, C.A. (2003). OpenSHA: A developing community-modeling environment for seismic hazard analysis. *Seismological Research Letters*, 74(4), 406–419.

Gahalaut, V.K. and Kundu, B. (2012). Possible influence of subducting ridges on the Himalayan arc and on the ruptures of great and major Himalayan earthquakes. *Gondwana Research*, 21(4), 1080–1088.

Galetzka, J., Melgar, D., Genrich, J.F., Geng, J., Owen, S., Lindsey, E.O., Xu, X., Bock, Y., Avouac, J.-P., Adhikari, L.B. et al. (2015). Slip pulse and resonance of the Kathmandu basin during the 2015 Gorkha earthquake, Nepal. *Science*, 349(6252), 1091–1095.

Ghazoui, Z., Bertrand, S., Vanneste, K., Yokoyama, Y., Nomade, J., Gajurel, A.P., van der Beek, P.A. (2019). Potentially large post-1505 AD earthquakes in western Nepal revealed by a lake sediment record. *Nature Communications*, 10(1), 2258.

Hanks, T.C. and Kanamori, H. (1979). A moment magnitude scale. *Journal of Geophysical Research: Solid Earth*, 84(B5), 2348–2350.

Hetényi, G., Cattin, R., Berthet, T., Le Moigne, N., Chophel, J., Lechmann, S., Hammer, P., Drukpa, D., Sapkota, D.S., Gautier, S. (2016). Segmentation of the Himalayas as revealed by arc-parallel gravity anomalies. *Scientific Reports*, 6(1), 33866.

Hubbard, J., Almeida, R., Foster, A., Sapkota, S.N., Bürgi, P., Tapponnier, P. (2016). Structural segmentation controlled the 2015 M w 7.8 Gorkha earthquake rupture in Nepal. *Geology*, 44(8), 639–642.

Jackson, M. and Bilham, R. (1994). Constraints on Himalayan deformation inferred from vertical velocity fields in Nepal and Tibet. *Journal of Geophysical Research: Solid Earth*, 99(B7), 13897–13912.

Kanamori, H. and Brodsky, E.E. (2004). The physics of earthquakes. *Reports on Progress in Physics*, 67, 1429–1496.

Kaneko, Y., Avouac, J.-P., Lapusta, N. (2010). Towards inferring earthquake patterns from geodetic observations of interseismic coupling. *Nature Geoscience*, 3(5), 363–369.

Kumar, S., Wesnousky, S.G., Jayangondaperumal, R., Nakata, T., Kumahara, Y., Singh, V. (2010). Paleoseismological evidence of surface faulting along the northeastern Himalayan front, India: Timing, size, and spatial extent of great earthquakes. *Journal of Geophysical Research*, 115(B12), B12422.

Lay, T. (2015). The surge of great earthquakes from 2004 to 2014. *Earth and Planetary Science Letters*, 409, 133–146.

Lindsey, E.O., Almeida, R., Mallick, R., Hubbard, J., Bradley, K., Tsang, L.L.H., Liu, Y., Burgmann, R., Hill, E.M. (2018). Structural control on downdip locking extent of the Himalayan megathrust. *Journal of Geophysical Research: Solid Earth*, 123(6), 5265–5278.

Maréchal, A., Mazzotti, S., Cattin, R., Cazes, G., Vernant, P., Drukpa, D., Thinley, K., Tarayoun, A., Le Roux-Mallouf, R., Bahadur Thapa, B. et al. (2016). Evidence of interseismic coupling variations along the Bhutan Himalayan arc from new GPS data. *Geophysical Research Letters*, 43(24), 12,399–12,406.

Métois, M., Vigny, C., Socquet, A. (2016). Interseismic coupling, megathrust earthquakes and seismic swarms along the Chilean Subduction Zone (38°–18°S). *Pure and Applied Geophysics*, 173(5), 1431–1449.

Michel, S., Avouac, J.-P., Lapusta, N., Jiang, J. (2017). Pulse-like partial ruptures and high-frequency radiation at creeping-locked transition during megathrust earthquakes. *Geophysical Research Letters*, 44(16), 8345–8351.

Michel, S., Avouac, J., Jolivet, R., Wang, L. (2018). Seismic and aseismic moment budget and implication for the seismic potential of the Parkfield segment of the San Andreas fault. *Bulletin of the Seismological Society of America*, 108(1), 19–38.

Michel, S., Jolivet, R., Rollins, C., Jara, J., Dal Zilio, L. (2021). Seismogenic potential of the main Himalayan thrust constrained by coupling segmentation and earthquake scaling. *Geophysical Research Letters*, 48(13), 1–10.

Mori, T., Shigefuji, M., Bijukchhen, S., Kanno, T., Takai, N. (2020). Ground motion prediction equation for the Kathmandu Valley, Nepal based on strong motion records during the 2015 Gorkha Nepal earthquake sequence. *Soil Dynamics and Earthquake Engineering*, 135, 106208.

Mugnier, J.L., Jouanne, F., Bhattarai, R., Cortes-Aranda, J., Gajurel, A.P., Leturmy, P., Robert, X., Upreti, B., Vassallo, R. (2017). Segmentation of the Himalayan megathrust around the Gorkha earthquake (25 April 2015) in Nepal. *Journal of Asian Earth Sciences*, 141(April 2015), 236–252.

Nhemafuki, G.R. (2008). Effect of local site condition on seismic ground response in Kathmandu Valley. Khwopa Engineering College, Purbanchal University, Gothgaun.

Ordaz, M., Martinelli, F., Aguilar, A., Arboleda, J., Meletti, C., D'Amico, V. (2015). CRISIS2015. Program for Computing Seismic Hazard. Instituto de Ingeniería. Universidad Nacional Autónoma de México, Mexico City, Mexico. Instituto de Ingenieria, UNAM.

Pagani, M., Monelli, D., Weatherill, G., Danciu, L., Crowley, H., Silva, V., Henshaw, P., Butler, L., Nastasi, M., Panzeri, L. et al. (2014). OpenQuake Engine: An open hazard (and risk) software for the global earthquake model. *Seismological Research Letters*, 85(3), 692–702.

Pandey, M.R. and Molnar, P. (1988). The distribution of intensity of the Bihar-Nepal earthquake of 15 January 1934 and bounds on the extent of the rupture zone. *Journal of Nepal Geological Society*, 5(1), 22–44.

Pandey, M.R., Tandukar, R.P., Avouac, J.P., Lavé, J., Massot, J.P. (1995). Interseismic strain accumulation on the Himalayan crustal ramp (Nepal). *Geophysical Research Letters*, 22(7), 751–754.

Pandey, B., Jakka, R.S., Kumar, A. (2016). Influence of local site conditions on strong ground motion characteristics at Tarai region of Uttarakhand, India. *Natural Hazards*, 81(2), 1073–1089.

Pudi, R., Joshi, S., Martha, T.R., Upadhyay, R., Pant, C.C. (2021). A comprehensive site response and site classification of the Garhwal-Kumaun Himalaya, Central Seismic Gap (CSG), India. *Journal of Earthquake Engineering*, 26(13).

Rana, B.S. (1935). *Nepal Ko Maha Bhukampa (The Great Earthquake of Nepal)*. Jorganesh Press, Kathmandu.

Reid, H.F. (1910). The California Earthquake of April 18, 1906, Report of the State Earthquake Investigation Commission, The Mechanism of the Earthquake. Nature (Vol. 2). Washington, DC.

Rollins, C. and Avouac, J. (2019). A geodesy- and seismicity-based local earthquake likelihood model for Central Los Angeles. *Geophysical Research Letters*, 46(6), 3153–3162.

Sapkota, S.N., Bollinger, L., Klinger, Y., Tapponnier, P., Gaudemer, Y., Tiwari, D. (2013). Primary surface ruptures of the great Himalayan earthquakes in 1934 and 1255. *Nature Geoscience*, 6(1), 71–76.

Savage, J.C. (1983). A dislocation model of strain accumulation and release at a subduction zone. *Journal of Geophysical Research: Solid Earth*, 88(B6), 4984–4996.

Shaw, B.E., Milner, K.R., Field, E.H., Richards-Dinger, K., Gilchrist, J.J., Dieterich, J.H., Jordan, T.H. (2018). A physics-based earthquake simulator replicates seismic hazard statistics across California. *Science Advances*, 4(8), eaau0688.

Stevens, V.L. and Avouac, J.-P. (2015). Interseismic coupling on the main Himalayan thrust. *Geophysical Research Letters*, 42(14), 5828–5837.

Stevens, V.L. and Avouac, J.-P. (2016). Millenary Mw > 9.0 earthquakes required by geodetic strain in the Himalaya. *Geophysical Research Letters*, 43, 1118–1123.

Stevens, V.L. and Avouac, J.-P. (2021). On the relationship between strain rate and seismicity in the India–Asia collision zone: Implications for probabilistic seismic hazard. *Geophysical Journal International*, 226(1), 220–245.

Stevens, V.L., Shrestha, S.N., Maharjan, D.K. (2018). Probabilistic seismic hazard assessment of Nepal. *Bulletin of the Seismological Society of America*, 108(6), 3488–3510.

Stevens, V.L., De Risi, R., Le Roux-Mallouf, R., Drukpa, D., Hetényi, G. (2020). Seismic hazard and risk in Bhutan. *Natural Hazards*, 104(3), 2339–2367.

Takai, N., Shigefuji, M., Rajaure, S., Bijukchhen, S., Ichiyanagi, M., Dhital, M.R., Sasatani, T. (2016). Strong ground motion in the Kathmandu Valley during the 2015 Gorkha, Nepal, earthquake. *Earth, Planets and Space*, 68(1), 10.

Tallett-Williams, S., Gosh, B., Wilkinson, S., Fenton, C., Burton, P., Whitworth, M., Datla, S., Franco, G., Trieu, A., Dejong, M. et al. (2016). Site amplification in the Kathmandu Valley during the 2015 M7.6 Gorkha, Nepal earthquake. *Bulletin of Earthquake Engineering*, 14(12), 3301–3315.

Valdiya, K.S. (1976). Himalyan transverse faults and folds and their parallelism with subsurface structures of North Indian plains: Discussion. *Tectonophysics*, 32(3–4), 353–386.

Wald, D.J. and Allen, T.I. (2007). Topographic slope as a proxy for seismic site conditions and amplification. *Bulletin of the Seismological Society of America*, 97(5), 1379–1395.

Wu, W., Liu, J., Guo, L., Deng, Z. (2020). Methodology and assessment of proxy-based Vs30 estimation in Sichuan Province, China. *International Journal of Disaster Risk Science*, 11(1), 133–144.

Ye, L., Lay, T., Kanamori, H., Rivera, L. (2016). Rupture characteristics of major and great ($M_w \geq 7.0$) megathrust earthquakes from 1990 to 2015: 2. Depth dependence. *Journal of Geophysical Research: Solid Earth*, 121(2), 845–863.

Zhao, J.X., Zhang, J., Asano, A., Ohno, Y., Oouchi, T., Takahashi, T., Ogawa, H., Irikura, K., Thio, H.K., Somerville, P.G. et al. (2006). Attenuation relations of strong ground motion in Japan using site classification based on predominant period. *Bulletin of the Seismological Society of America*, 96(3), 898–913.

Part 3

Focus

7

Recent and Present Deformation of the Western Himalaya

François JOUANNE[1], Jean-Louis MUGNIER[1], Riccardo VASSALLO[1], Naveed MUNAWAR[2], Awais AHMED[2], Adnan Alam AWAN[2], Manzoor A. MALIK[3] and Ramperu JAYANGONDAPERUMAL[4]

[1] *Savoie Mont Blanc University, Le Bourget-du-Lac, France*

[2] *Geological Survey of Pakistan, Islamabad, Pakistan*

[3] *University of Jammu, India*

[4] *Wadia Institute of Himalayan Geology, Dehra Dun, India*

7.1. Introduction

The northwestern part of the Himalaya, located west of longitude 76°E, shows structural variations compared to the central Himalaya (Figure 7.1 and Volume 1 – Chapters 3 and 4). Three regions constitute this area (i) the eastern zone, including the Kashmir basin, the Pir Panjal range and the Jammu region, (ii) the western zone, including the Salt Ranges, the Potwar Plateau and the Peshawar basin and (iii) the Nanga Parbat syntaxis.

Himalaya, Dynamics of a Giant 3,
coordinated by Rodolphe CATTIN and Jean-Luc EPARD.

This chapter is a focus on the thrust tectonics and the difference between the western and eastern parts of this area. To do so, we detail the role of lithological and structural inheritance that led to the contrast between the long-term evolution of the thick-skinned Jammu–Kashmir zone thrust system and the thin-skinned Pakistan system. Next, we present the main features of the MWT (Medlicott-Wadia Thrust), subject to numerous out-of-sequence reactivations during the Quaternary and affected by a magnitude Mw 7.6 earthquake in 2005. Finally, we discuss the role of asperities and dislocations in the present-day deformation observed in the western Himalaya.

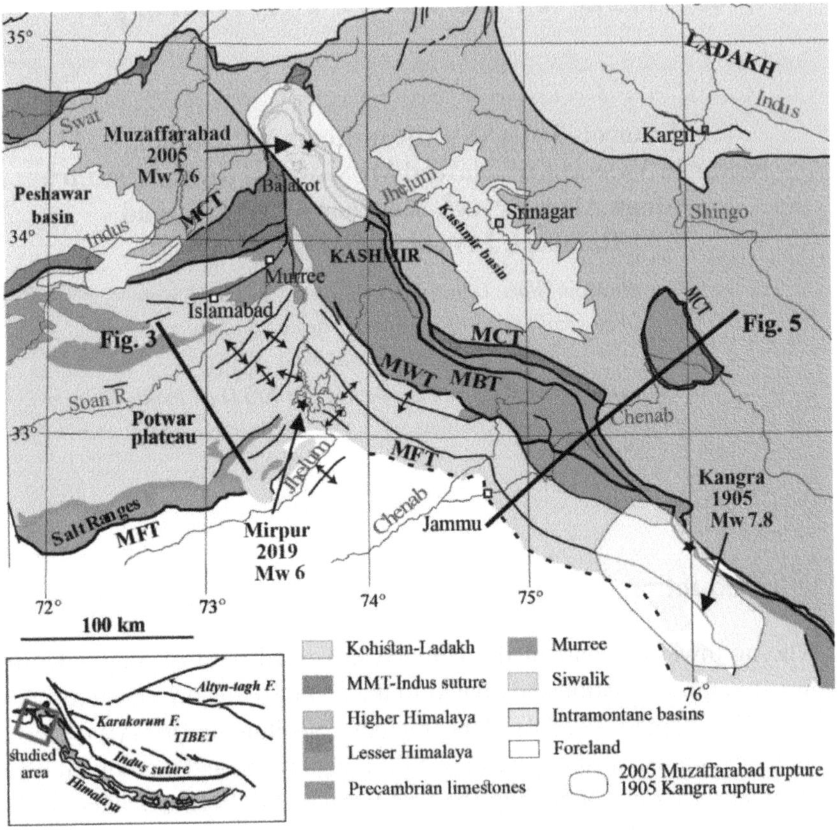

Figure 7.1. *Simplified structural map of the northwestern Himalayan front in the northwestern syntaxis region (adapted from Vassallo et al. 2015). For a color version of this figure, see www.iste.co.uk/cattin/himalaya3.zip*

7.2. Structural styles and tectonic prism model

7.2.1. *The main structural zones*

The western Himalayan syntaxis marks the boundary between the western and eastern thrust of the northwestern Himalaya. This syntaxis is extremely complex. In the northernmost part, the ongoing upward Nangat Parbat granite constitutes the heart of the syntaxis (Butler 2019). In the outermost part, this boundary corresponds to the Jhelum fault (Burbank and Beck 1989). Its kinematics is still debated, with significant rotations on either side of the fault and a suggested strike-slip component right lateral or left lateral, depending on the areas, periods and authors.

The presence of serpentine along the Jhelum fault zone suggests a tectonic structure pre-dating the Himalayan orogeny. On either side of this fault, we can observe a marked change in the lithology at the base of the series involved in Himalayan tectonics. This change suggests a pre-Himalayan paleogeography inheritance. The tectonics of the northwestern Himalaya thus evolves between two types of orogenic wedges with different basal frictions:

– in the east a prism with high basal friction;

– in the west a prism with low basal friction, for which the décollement level gradually flows thanks to the Eocambrian evaporites.

7.2.2. *The Himalaya in the west of the syntax*

The western part of Pakistan is a classic example (Yeats and Lawrence 1984) of a detached prism that evolves as a "thin-skinned tectonic" structure with several basins transported over 7 Ma in the Salt Ranges and Kohat fold and thrust belt.

The Salt Range and Potwar plateaus in Pakistan are composed of four units covering the rigid Indian craton (Grelaud et al. 2002):

– the Neoproterozoic Salt Ranges Formation with its basal décollement level;

– the Cambrian-Eocene shell sequence;

– the Rawalpindi Group;

– the Siwalik Group (Neogene molasse).

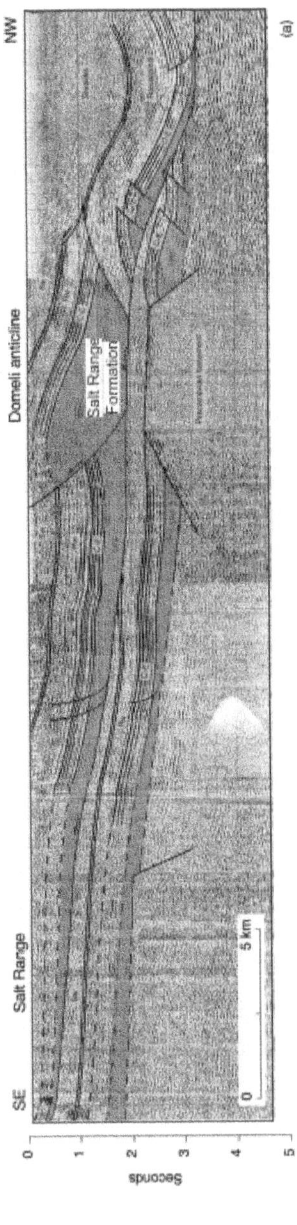

Figure 7.2. *Seismic profile illustrating the sedimentary series and structures of the Potwar Plateau (in pink the salt level of the Salt Ranges) (adapted from Grelaud et al. 2002). The salt levels show a lack of continuity related either to a diapiric flow, an offset by normal faulting affecting the bedrock or a non-continuous initial deposition. For a color version of this figure, see www.iste.co.uk/cattin/himalaya3.zip*

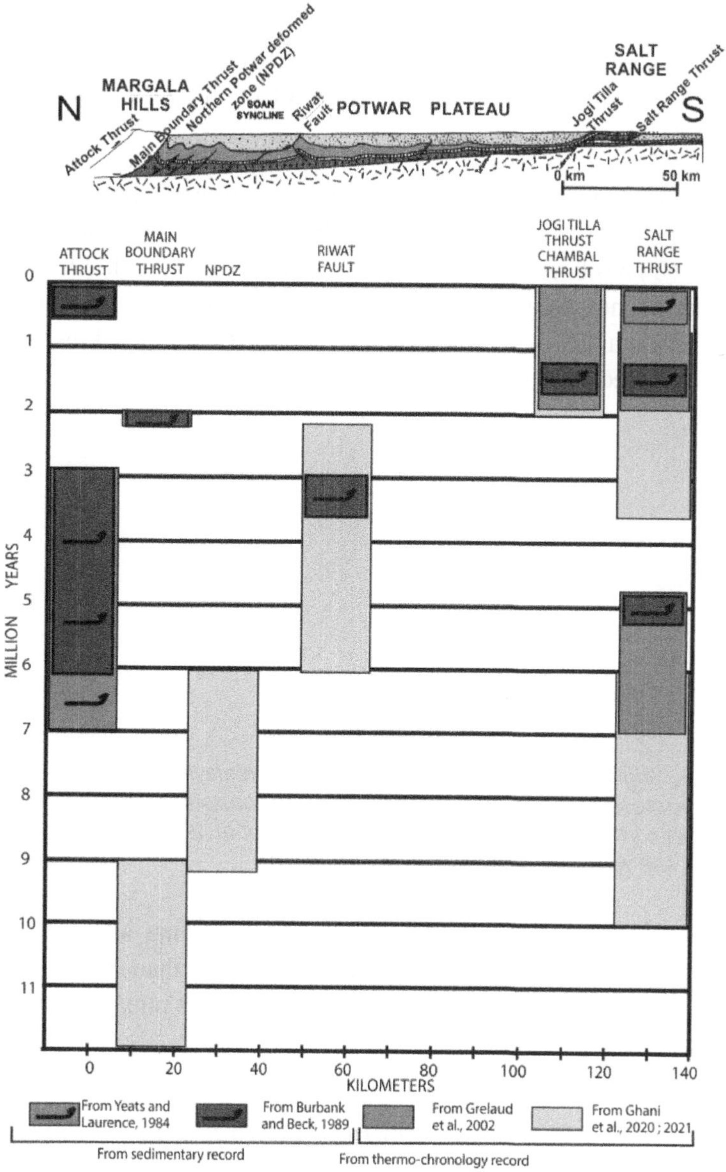

Figure 7.3. *Periods of thrust activity determined from sedimentary message (arrows) or thermochronology (after Yeats and Lawrence 1984; Burbank and Beck 1989; Grelaud et al. 2002; Ghani et al. 2020, 2021) along a synthetic section across the Salt Ranges (after Baker et al. 1988). For a color version of this figure, see www.iste.co.uk/cattin/himalaya3.zip*

Ante-Himalayan normal faults affect the bedrock, with geometric and lithologic asperities associated with ramps and change in thickness of the salt – local disappearance or stuffing at the heart of the anticlines, respectively (Figure 7.2).

The frontal thrust is outcropping at the Salt Range Thrust (SRT), where extensional structures can be observed (Butler et al. 1987). The Main Frontal Thrust (MFT) has been active for at least 10 Ma (Ghani et al. 2021). However its activity is not continuous, with a complex sequence of fault activity (Figure 7.3) characterizing the orogenic wedge (Yeats and Lawrence 1984; Burbank and Beck 1989; Grelaud et al. 2002; Ghani et al. 2021).

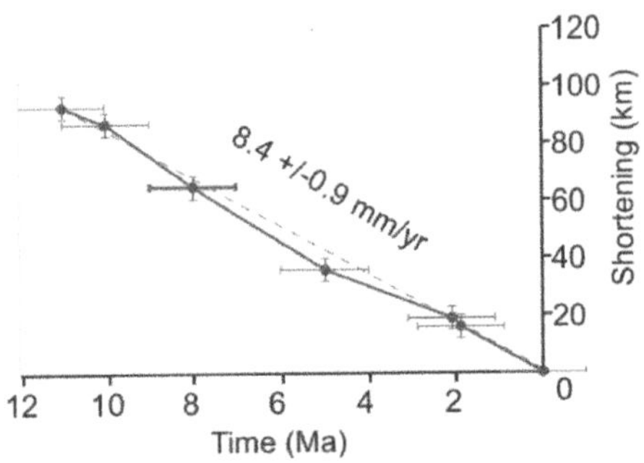

Figure 7.4. *Shortening over time in the western Himalayan syntaxis estimated from six geologic cross-sections. These shortenings are consistent with a shortening rate of 8.4 ± 0.9 mm.yr^{-1} (Synthesis after Cortés-Aranda et al. 2017). For a color version of this figure, see www.iste.co.uk/cattin/himalaya3.zip*

Despite this complex activity, the rate of shortening affecting the entire western syntaxis has been relatively constant for more than 10 Ma (Figures 7.3 and 7.4) with an average velocity of approximately 8 mm/yr (synthesis after Cortés-Aranda et al. 2017).

7.2.3. *The Himalaya of the Jammu–Kashmir area*

"Thick-skin tectonics" characterizes the Jammu–Kashmir Himalaya (Srinivasan and Khar 1996) (Figure 7.5). The frontal structure at the boundary of the foreland plain is formed by an anticline associated with either a detachment fold or a triangular zone involving deep-seated tectonic slices and retrothrusting. The frontal thrust does not propagate to the surface but remains blind. The bedrock is involved even in very outer areas. The depth of the décollement level there exceeds 7 km. This detachment develops in many cases at the base or along localized stratigraphic contacts in the Precambrian limestones (Sirban limestones). The upper décollement level deepens abruptly to more than 25 km. Above a crustal ramp, a rapid uplift favors the development of tectonic windows such as the Kishtwar window (Figure 7.5).

7.2.4. A structural evolution in agreement with the tectonic prism model

The tectonic critical prism model (Dahlen 1990) is based on simple mechanical and geometrical assumptions (Figure 7.6). However, it accounts for the main characteristics of the northwestern Himalaya over long timescale (> 10 Ma):

1) The lithology and rheology of the décollement level have a key role in the thickening and frontal thrust propagation. The eastern part is characterized by a slight rheological contrast between the transported tectonic prism and the carbonates of the décollement level. The thickness of the prism and its average surface slope are high, with an elevation increase of approximately 3,000 m in 70 km, that is, an average gradient of 4%. Westward, the low mechanical strength of the evaporites of the Neoproterozoic décollement level favors a frontal propagation far forward with an average surface slope < 1%. This behavior induces an increase of 1,000 m in 125 km and causes SRT activity as early as 10 Ma (Figure 7.3). It also favors the synchronous development of piggy-back basins above the décollement, such as Potwar and Peshawar basins.

2) The sequence of thrust activity in the prism is related to interactions between tectonics and surface processes (see Volume 3 – Chapter 2). In the west, sedimentation affects the prims dynamics (Leturmy et al. 2000). Flexure of the foreland basin and deposition of syn-orogenic sediments enhance the prismatic shape of the deposits above the potential décollement level of the Neoproterozoic salt formations. These processes allow a critical prism to be reached very early, and the frontal thrust (SRT) has been active for at least 10 Ma (Figure 7.3). The thrusts transported by this basal detachment have been propagating outward from the Main Boundary Thrust (MBT) to the Jogi Tilla Thrust for more than 12 Ma in a sequence characteristic of a prism (Figure 7.3). In the east, out-of-sequence reactivations are required to maintain thickening and compensate for erosion (Dahlen 1990). This out-of-sequence deformation is particularly well expressed along the MWT during the Quaternary and most recent period.

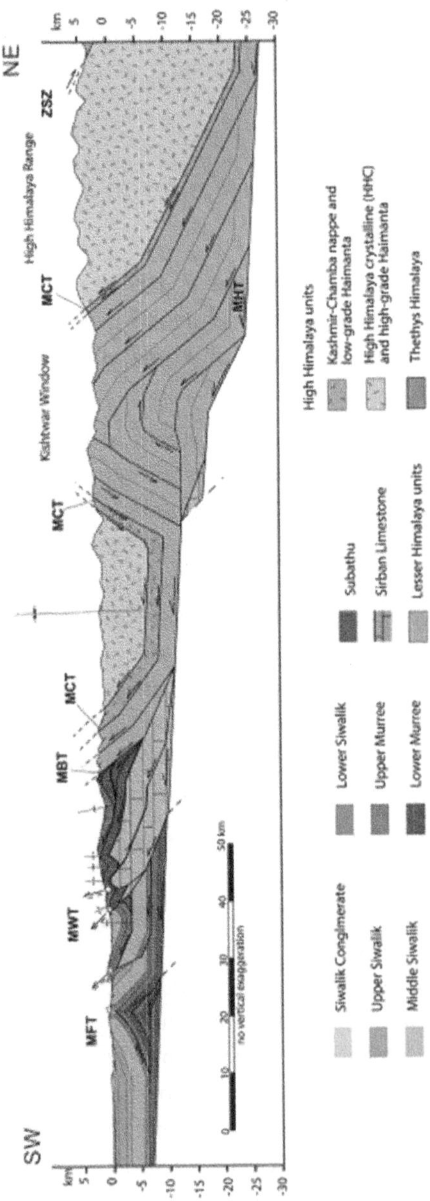

Figure 7.5. *Cross-section through the Jammu–Kashmir Himalaya (after Srinivasan and Khar 1996; Gavillot et al. 2018). MFT: Main Frontal Thrust; MWT: Medicott-Wadia Thrust; MBT: Main Boundary Thrust; MCT: Main Central Thrust; ZSZ: Zanskar Shear Zone. For a color version of this figure, see www.iste.co.uk/cattin/himalaya3.zip*

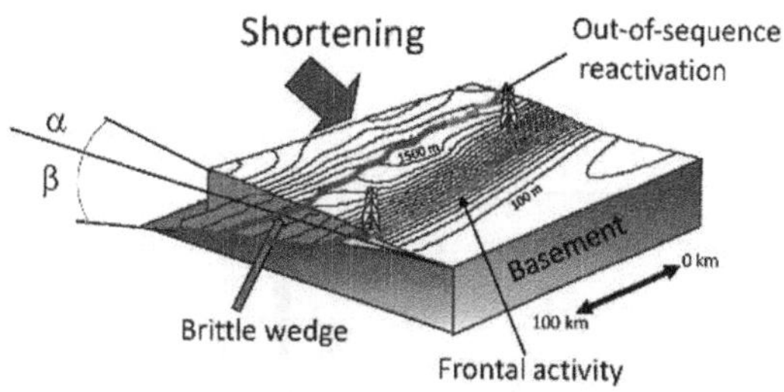

Figure 7.6. *Block diagram illustrating the critical prism model. α and β are the average slopes of the topographic surface and the décollement level, respectively (adapted from Dahlen 1990). For a color version of this figure, see www.iste.co.uk/cattin/himalaya3.zip*

7.3. Out-of-sequence deformation in the western Himalayan syntax

Out-of-sequence deformation characterizes the northwest Himalayan region from the scale of thousands of years to historical seismicity. This deformation mode has been revealed by combining morphotectonics, paleoseismology and seismology approaches.

7.3.1. *Quaternary out-of-sequence activity of the NW Himalayan thrusts*

A system of sub-parallel thrusts extends for several hundred kilometers from the syntax area in the northwest to the Kangra region in the southeast (Figure 7.7). In the Jammu–Kashmir region, three major Quaternary thrusts have shaped the present-day landform by uplifting their respective hanging wall and creating hundreds of meters high tectonic scarps (Figure 7.8) (Vassallo et al. 2015).

The Main Boundary Thrust (MBT), located a few tens of kilometers from the front, is the innermost thrust fault. Its trace is associated with a kilometer-long morphological step that suggests significant cumulative deformation. Although Burbank and Johnson (1983) proposed a Quaternary activity of this thrust, the terraces and alluvial fans that overlie the fault are not deformed. Their dating implies that the MBT has been inactive for at least 30,000 years (Vassallo et al. 2015). This result means that the relief around the MBT is associated with past tectonic activity, overlaid by differential erosion related to the contrast of lithologies juxtaposed by the fault.

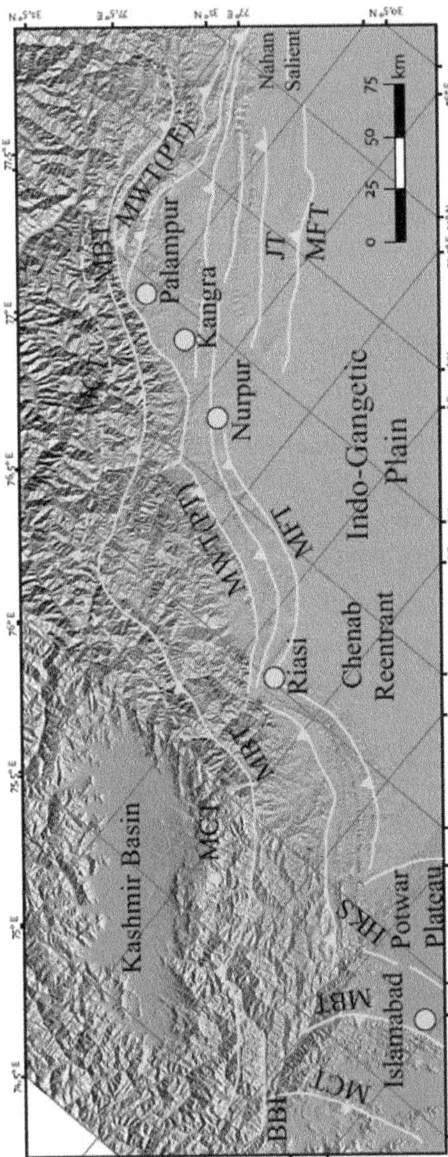

Figure 7.7. *Tectonic setting of the Sub-Himalayan and adjacent regions to the east of the West Himalayan syntaxis. The active faults of the Bilaspur Thrust, Palampur Thrust, Riasi Thrust, Balakot-Bagh Fault are collectively referred to as the Medicott-Wadia Thrust (MWT) (adapted from Cortés-Aranda et al. (2017). For a color version of this figure, see www.iste.co.uk/cattin/himalaya3.zip*

The Medlicott-Wadia Thrust (MWT), approximately 200 km long, lies between the MBT and the Himalayan front (Thakur et al. 2010). In the Riasi area, where the fault has a reentrant strike to the inner range, the MWT accommodates the thrusting of Precambrian carbonates over the Neogene molasse deposits of the Siwaliks. In this area, this fault affects Quaternary deposits and morphologies. An alluvial fan of 14 ± 2 ka dated by OSL is deformed by the action of several branches of the MWT (Figure 7.8) that connect at depth to a approximately 45° northeastward dipping fault (Vignon et al. 2017). These faults have produced a 150-m tectonic scarp since the fan is inactive. In addition, a succession of seven fluvial terraces record incision by the Chenab River that reaches 370 m since 36,000 years (Vignon et al. 2017). This morphological analysis suggests a shortening across this out-of-sequence thrust fault system of approximately 11 mm/yr over the last 14,000 years (Vassallo et al. 2015; Gavillot et al. 2016). Therefore, in this part of the range, the MWT alone accommodates most of the India-Eurasia convergence estimated to 14–15 mm/yr from GNSS measurements (Jouanne et al. 2020).

The Frontal Anticline (FA) is the outermost active Himalayan structure. This anticline is associated with the Main Frontal Thrust (MFT), which is a blind thrust in the region. The shortening rate across this structure is assessed from a morpho-chronological analysis of terraces folded by the progressive uplift of the anticline (Vassallo et al. 2015). Using the anticline excess area method (Hossack 1979), this geomorphological study is then coupled with an estimate of the major décollement depth obtained from seismic profiles (Raiverman et al. 1994; Srinivasan and Khar 1996). Together, these approaches suggest a shortening rate of 9 ± 3.2 mm/yr accomoadated along the MFT over the last 24,000 years (Vassallo et al. 2015).

Further southeast, in the Kangra region, the tectonic structures are located in a long-wavelength reentrant (~100 km). In this region, the active deformation zone reaches a width of 90 km, which has no equivalent in the Himalayan range. Thrust faults associated with tectonic scarps are multiple with a limited geographic extent (Thakur et al. 2020). Quaternary deformation is thus distributed over several ramps that root at depth on the main décollement (Powers et al. 1998). The Palampur Thrust (Figure 7.7) is the innermost fault that deforms the surficial formations. Scarps of a few meters characterize the alluvial terraces that are intersected by this fault. Their Holocene dating gives an estimated slip rate along this fault of approximately

1 mm/yr, which represents approximately 10% of the total shortening in the Kangra reentrant (Cortés-Aranda et al. 2018).

In the northwestern Himalaya, active deformation is thus distributed across several crustal ramps, acting along localized out-of-sequence structures up to 90 km from the mountain front. This accommodation of the deformation along internal faults is probably one of this region's most unique tectonic aspects compared to the other parts of the Himalaya. The deformation style differs between the northern and southern parts of this region. In the north, two major structures accommodate the shortening, whereas deformation is spread over several more minor thrust faults in the south.

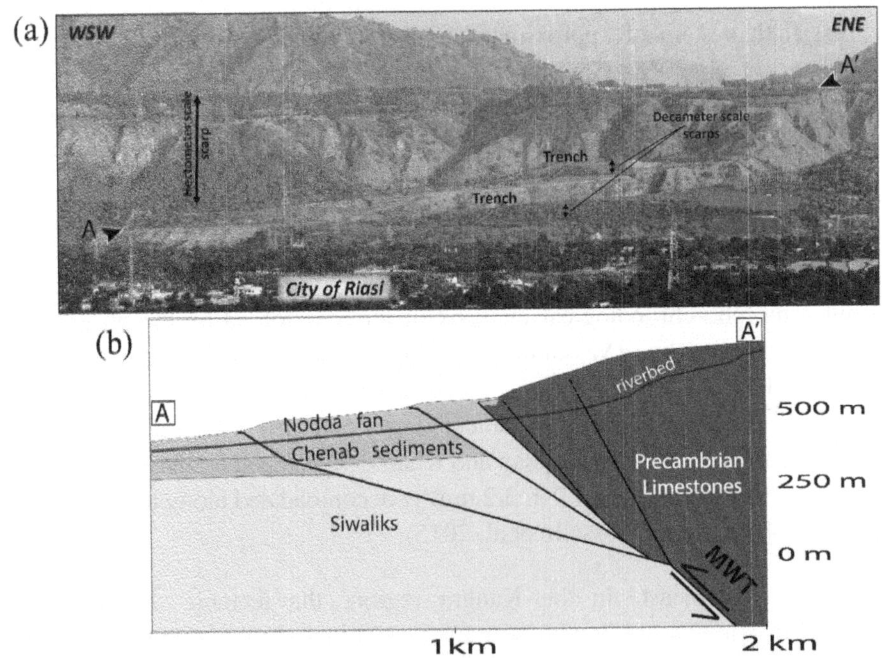

Figure 7.8. *Cumulative out-of-sequence deformation revealed by morphology on the MWT. (a) Panoramic view of the hectometer-scale tectonic scarp produced by the MWT in the Riasi area. In the Nodda River alluvial fan, the MWT has split into several fault branches associated with decameter-scale escarpments. The two outermost fault branches and the scarp related to these faults are decameters in scale. Trench sites (Figure 7.9) are located on these two fault branches and (b) A–A' cross-section of the scarp showing major fault branches (modified from Vassallo et al. 2015). For a color version of this figure, see www.iste.co.uk/cattin/himalaya3.zip*

7.3.2. *Out-of-sequence seismological and paleoseismological activity in the NW Himalaya*

The still ongoing out-of-sequence deformation controls the seismic activity in the region. Indeed, despite a five-century seismic quiescence throughout Jammu and Kashmir, the MWT again stands out as the most important regional source of seismicity in the Holocene. Two paleoseismological trenches across the fault zone at Riasi show at least five earthquakes over the last ~3,500 years (Vassallo et al. 2020). Each event is characterized by several meters of co-seismic displacement, suggesting the occurrence of high magnitude earthquakes along this fault (Figure 7.9). Furthermore, based on several geological and historical clues, the MWT is the most probable fault that produced the last large earthquake in Kashmir in 1555 AD.

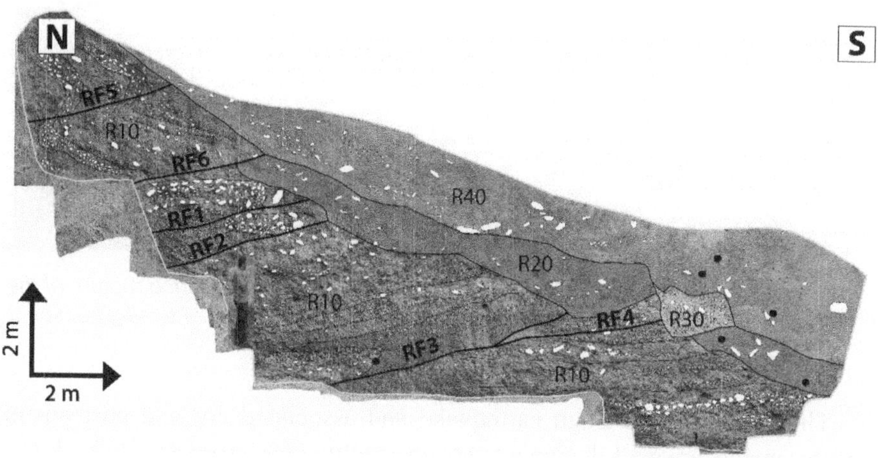

Figure 7.9. *Riasi trench through the MWT. Mosaic view of the eastern wall of the trench through the "Rain fault" branch with the interpretative log indicating a superposition of six faults. The black circles give the location of the dated samples. For a color version of this figure, see www.iste.co.uk/cattin/himalaya3.zip*

The only instrumental earthquake with surface ruptures in the Himalaya is the Balakot-Bagh earthquake, which occurred in this region in 2005. This out-of-sequence Mw 7.6 earthquake was one of the deadliest in history, with 80,000 casualties. This rupture produced a surface slip along the fault plane up to 7 m generating a co-seismic scarp several meters high (Kaneda et al. 2008). Based on a paleoseismological study, Kondo et al. (2008) propose an earthquake of equivalent magnitude during the last ~2,000 years in the same region. The 2005 event can be viewed as the latest episode of a long-term activity that results in tectonic scarps several tens of meters high as observed along the MWT at Riasi (Figure 7.10).

Figure 7.10. *Cumulative tectonic scarp at Muzaffarabad and surface rupture of the 2005 Balakot-Bagh earthquake (red line). For a color version of this figure, see www.iste.co.uk/cattin/himalaya3.zip*

The 2005 Balakot-Bagh earthquake and associated co- and post-seismic displacements are well-documented from satellite interferometry methods (e.g. Pathier et al. 2006, and Figure 7.11A) and geodetic observations (Jouanne et al. 2011). Numerical modeling suggests a coseismic motion along a 30° dip ramp through 15 km of crust. The rupture, initiated at the base of the ramp, reaches a maximum amplitude of 10 m near the surface. The post-seismic motion takes place along a shallow NE-dipping fault (Figure 7.11B). The attenuation of displacement over time is consistent with rate-strengthening frictional sliding.

The northwestern Himalaya is thus an outstanding example of out-of-sequence activity on multiple timescales. In particular, this region provides an excellent tectonic example showing that out-of-sequence deformation can be related to earthquakes. It also demonstrates the relevance of approaches combining critical prism models with dynamic rupture models of a connected faults system (DeDontney and Hubbard 2012).

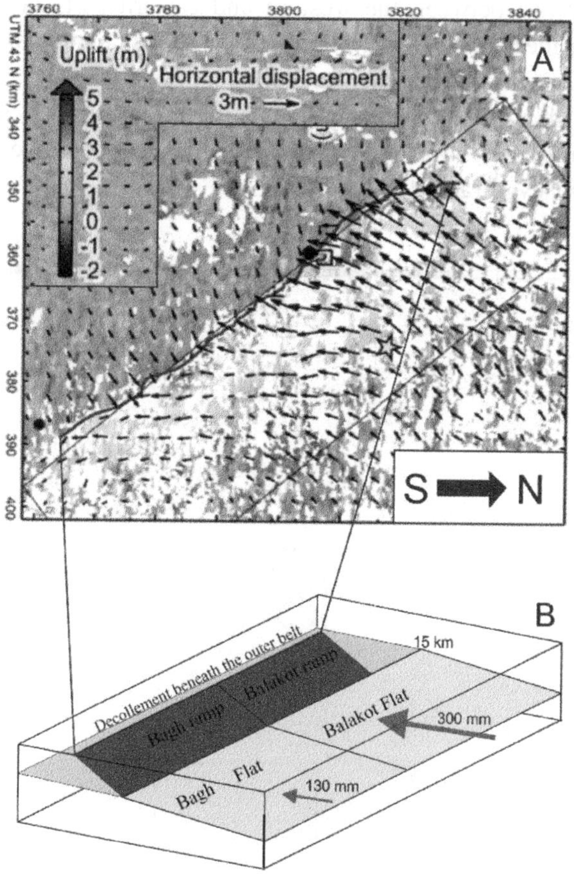

Figure 7.11. *(A) 3D displacement field of the topographic surface. Arrows and colors indicate horizontal and vertical displacements, respectively. Arrows are every 3.5 km. From north to south, the black circles represent the cities of Balakot, Muzaffarabad and Bagh. The red line shows the trace of the surface rupture (adapted from Pathier et al. 2006). (B) Depth geometry of the Balakot-Bagh fault, with ramp segments affected by co-seismic motion (red) and the 10° northeast-sloping décollement (yellow) affected by post-seismic slip. The décollement below the sub-Himalaya is shown in purple. The green arrows correspond to the post-seismic displacement during the nine months following the earthquake (adapted from Jouanne et al. 2011). For a color version of this figure, see www.iste.co.uk/cattin/himalaya3.zip*

7.4. Deformation associated with a ductile décollement, not always aseismic

The difference between the western and eastern parts of the study area is also underlined by the distribution of historical earthquakes. While the historical seismicity shows high-magnitude events in the Jammu–Kashmir Himalaya, further west high-magnitude earthquakes are not described, except may be an earthquake in 25 AD that would have destroyed the ancient city of Taxila (Ambraseys et al. 1975).

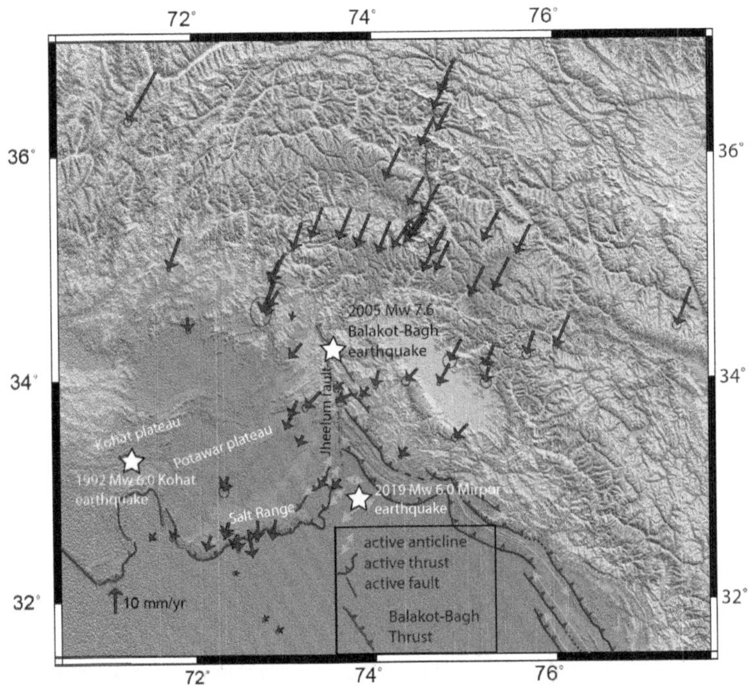

Figure 7.12. *GNSS-measured displacement velocity expressed relative to the assumed fixed India (Jouanne et al. 2020) and location of Kohat, Balakot-Bagh and Mirpur earthquakes that occurred in 1992, 2005 and 2019, respectively. For a color version of this figure, see www.iste.co.uk/cattin/himalaya3.zip*

7.4.1. *Seismic coupling assessment along the MHT*

Geodetic data acquired since 2001 show a very uniform displacement field in the Jammu and Kashmir Himalaya (Khan et al. 2008; Jouanne et al. 2020).

In contrast, in the Salt Ranges, rapid spatial variations in displacement velocity are observed (Figure 7.12).

Considering a geometry constrained by receiver function data, the coupling coefficient along the MHT is estimated from numerical modeling (Figure 7.13 and Jouanne et al. 2020).

In the Salt Ranges and under the Potwar basin, we observe rapid variations in coupling (Figure 7.13). This suggests the existence of weak coupling areas with an excellent quality of décollement and regions coupled with a poorer level of décollement. We can thus expect earthquakes of moderate magnitudes along these asperities and associated coupled zones.

7.4.2. *Occurrence of ruptures on asperities*

While the Kohat Plateau is associated with a good level of décollement formed by Eocambrian evaporites, the 1992 Mw 6.0 earthquake (Figures 7.12 and 7.14) suggests the existence of asperities potentially prone to rupture during earthquakes.

We, therefore, assume that these asperities of the Kohat Plateau and Potwar Plateau correspond to areas where the salt level does not exist or, more precisely, no longer exists. The horizontal and vertical flow of salt can affect the salt thickness along the detachment plane, leading to zones without salt, forming asperities, and areas with a thickened salt layer. This lateral flow partly controls deformation, as suggested by Jouanne et al. (2020) for the Salt Ranges, and can locally induce the formation of normal faults at the top of zones where the salt flows or has flowed (Butler et al. 1987). Furthermore, this process can also control post-seismic deformation south of the rupture zone, as a strong earthquake along the deep part of the MHT can accelerate salt flow when localized in the basement. The same phenomenon occurs along the western edge of the Salt Ranges and Potwar Plateau, the Kalabagh Fault, marked by diapirs and segments affected by moderate magnitude earthquakes with an estimated recurrence of 2,000 years (Abbas et al. in press).

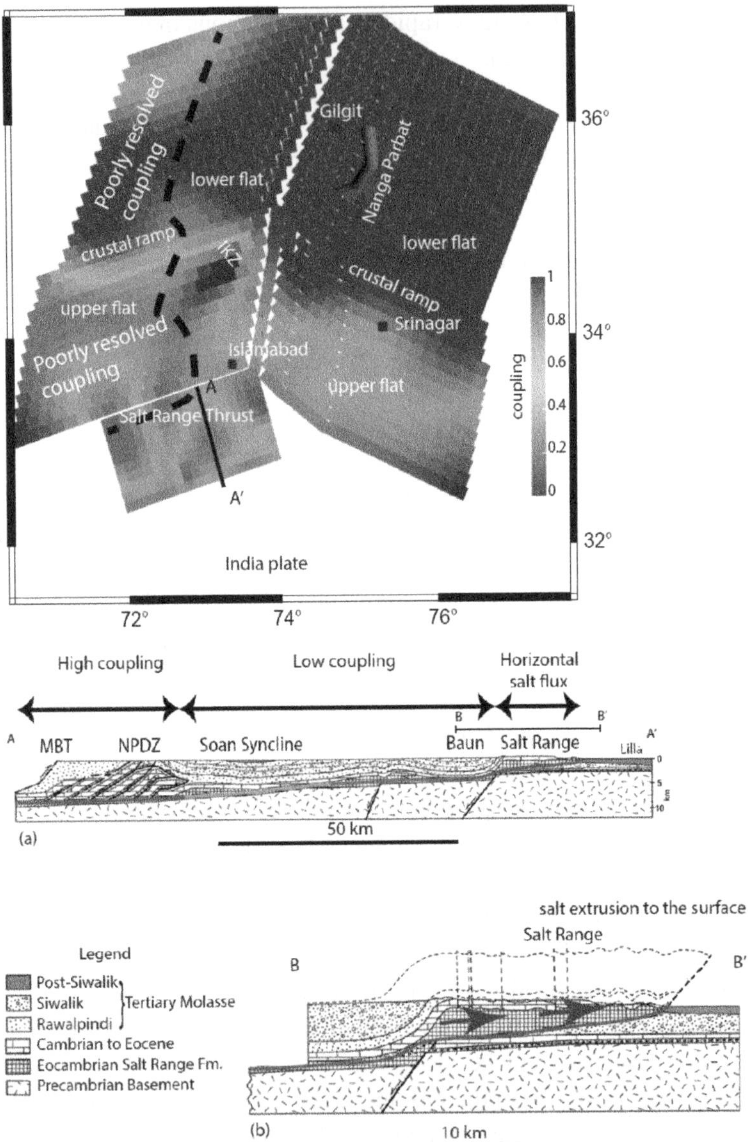

Figure 7.13. *Coupling assessment along the MHT from geodetic data (Jouanne et al. 2020) and evolution of coupling along a section across the Potwar Basin and Salt Ranges. For a color version of this figure, see www.iste.co.uk/cattin/himalaya3.zip*

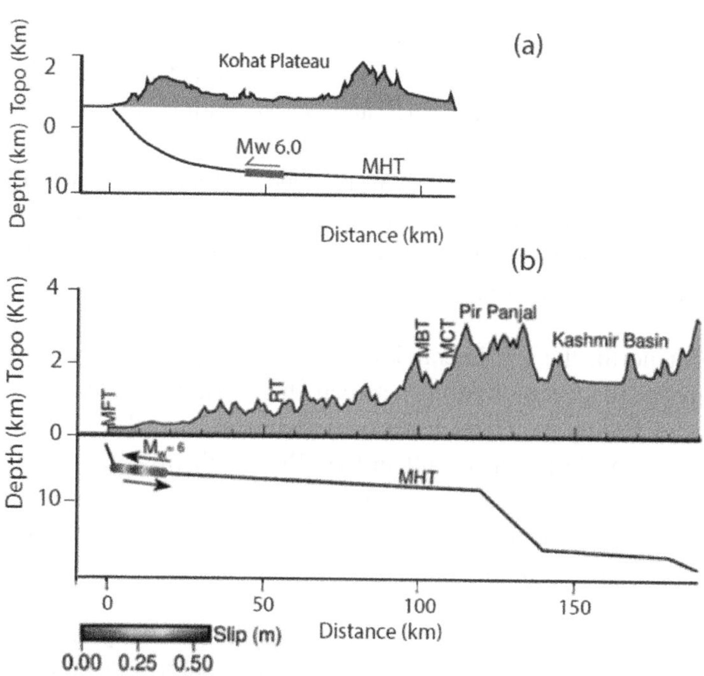

Figure 7.14. *Location of ruptured areas along the MHT in the Kohat (a) and Mirpur (b) earthquakes, respectively, according to Satyabala et al. (2012) and Sreejith et al. (2021). The location of these two magnitudes Mw 6.0 earthquakes is shown in Figure 7.12. For a color version of this figure, see www.iste.co.uk/cattin/himalaya3.zip*

Further north, still in the western part of the study area, basal thrusting continues into the basement, where a strong coupling is estimated. The MHT flattens northward, and the coupling becomes zero, suggesting ductile deformation (Figure 7.13). This horizontal ductile layer was the site of post-seismic deformation, probably related to afterslip following the Balakot-Bagh Mw 7.6 earthquake in 2005 (Jouanne et al. 2011; Wang and Fialko 2014). This earthquake, located along an out-of-sequence thrust as discussed above, is localized along a ramp that connects to the MHT at the brittle–ductile transition, which is marked by the change between zero coupling to the north and strong coupling to the south.

7.4.3. *Transition between thin-skin and thick-skin tectonics*

The MHT has a flat geometry in the Mangla Dam region east of the Salt Ranges. This area is a west–east transition zone between a shallow MHT with low coupling and a deep and fully coupled MHT (Jammu–Kashmir Himalaya). This transition region was recently affected by the 2019 Mw 6 Mirpur earthquake (Figures 7.12 and 7.14) studied by InSAR (Sreejith et al. 2021). This seismic event located along a 5-km deep horizontal thrust, which corresponds to the MHT. Due to human activity, the loss of coherence associated with agriculture did not allow the study of possible post-seismic deformation related to the salt level documented west of the earthquake-affected region.

Further north, the left-lateral Jheelum strike-slip fault accommodates the difference in deformation style between the western part marked by cover tectonics, a locally excellent level of detachment, and the Jammu–Kashmir Himalaya. Considering this fault in the GNSS velocity simulation (Jouanne et al. 2020) suggests an unlocked structure with a dextral strike-slip rate of 6 mm/yr, including near the surface. This creep would be explained by the serpentinite observed in the fault zone.

Further east, a deep MHT with a flat-ramp-flat geometry characterizes the Jammu–Kashmir Himalaya. The MHT exhibits a coupling distribution at depth similar to Nepal's (Bettinelli et al. 2006; Jouanne et al. 2017; Dal Zilio et al. 2020), with a fully coupled upper flat without short-wavelength variations, a partially coupled crustal ramp and an uncoupled lower flat.

7.5. Conclusion

The northwestern Himalaya provides a remarkable illustration of the long-term evolutionary characteristics of the tectonic critical prism model. The lithology and rheology of the décollement level play a key role in thickening distribution and frontal thrust propagation. A low mechanical strength detachment zone favors the development of piggy-back basins above the décollement, whereas the sequence of thrust activity in the prism is related to interactions between shortening, erosion and sedimentation.

Out-of-sequence deformation is particularly well expressed in the northwestern Himalaya. The active deformation in this region is distributed over several thrusts, from the front to 90 km inner part of the range. The most spectacular example is the MWT in the Riasi region, which is associated with Quaternary tectonic scarps over 100 m high and absorbs the majority of the India–Eurasia convergence. Historical and Holocene seismicity follows this pattern, with the 2005 Balakot-Bagh, 1905 Kangra and 1555 Great Kashmir earthquakes.

Finally, present-day deformation models suggest locked asperities over different spatial and probably temporal scales during seismic cycle. Even along a well-documented ductile detachment, such as the Salt Ranges, asperities prevent aseismic slip and can generate moderate magnitude earthquakes. The ductile and shallow décollement is rooted northward in the basement. The Himalayan thrust can be temporarily locked (stick-slip) and likely produce significant earthquakes when this décollement is localized in the basement.

7.6. References

Abbas, W., Zhang, J., Tsukamoto, S., Ali, S., Frechen, M., Reicherter, K. (in press). A Pleistocene-Holocene deformation and seismic history of the Kalabagh Fault in Pakistan using OSL and post-IR IRSL dating. *Quaternary International*.

Ambraseys, N., Lensen, G., Moinfar, M. (1975). The Pattan earthquake of 28 Dec., 1974. Reconnaissance Report Prepared for the Government of Pakistan and UNESCO, Paris.

Baker, D.M., Lillie, R.J., Yeats, R.S., Johnson, G.D., Yousuf, M., Zamin, A.S.H. (1988). Development of the Himalayan frontal thrust zone: Salt Range, Pakistan. *Geology*, 16, 3–7.

Bettinelli, P., Avouac, J.-P., Flouzat, M., Willis, P., Jouanne, F., Bollinger, L., Chitrakar, G.R. (2006). Interseismic deformation in the Himalaya of Central Nepal from combined GPS measurements and DORIS data. *Journal of Geodesy*, 80, 567–589.

Burbank, D.W. and Beck, R.A. (1989). Early Pliocene uplift of the Salt Range; temporal constraints on thrust wedge development, northwest Himalaya, Pakistan. In *Tectonics of the Western Himalaya*, Malinconico Jr., L.L. and Lillie, R.J. (eds). Geological Society of America Special Paper, McLean.

Burbank, D.W. and Johnson, G.D. (1983). The Late Cenozoic chronologic and strati-graphic development of the Kashmir intermontane basin, Northwestern Himalaya. *Palaeogeography, Palaeoclimatology, Palaeoecology*, 43, 205–235.

Butler, R.W. (2019). Tectonic evolution of the Himalayan syntaxes: The view from Nanga Parbat. *Geological Society Special Publications*, 483(1), 215–254.

Butler, R.W., Coward, M.P., Harwood, G.M., Knipe, R.J. (1987). Salt control on thrust geometry, structural style and gravitational collapse along the Himalayan mountain front in the Salt Range of northern Pakistan. *Dynamical Geology of Salt and Related Structures*, 399, 339–418.

Cortés-Aranda, J., Mugnier, J.-L., Vassallo, R., Jouanne, F., Carcaillet, J., Adnan, A. (2017). Holocene shortening rates and seismic hazard assessment for the frontal Potwar Plateau, NW Himalaya of Pakistan: Insights from 10Be concentrations on fluvial terraces in the Mahesian Anticline. *Quaternary International*, 462, 75–89.

Cortés-Aranda, J., Vassallo, R., Jomard, H., Pousse-Beltran, L., Astudillo, L., Mugnier, J.-L., Jouanne, F., Malik, M., Carcaillet, J. (2018). Late Quaternary out-of-sequence deformation in the innermost Kangra Reentrant, NW Himalaya of India: Seismic potential appraisal from 10Be dated fluvial terraces. *Journal of Asian Earth Sciences*, 158, 140–152.

Dahlen, F.A. (1990). Critical taper model of fold-and thrust belts and accretionary wedges. *Annual Review of Earth and Planetary Sciences*, 18, 55–99

Dal Zilio, L., Jolivet, R., van Dinther, Y. (2020). Segmentation of the main Himalayan thrust illuminated by Bayesian inference of interseismic coupling. *Geophysical Research Letters*, 47, e2019GL086424.

DeDontney, N. and Hubbard, J. (2012). Applying wedge theory to dynamic rupture modeling of fault junctions. *Bulletin of the Seismological Society of America*, 102(4), 1693–1711.

Gavillot, Y., Meigs, A.J, Yule, Y., Heennanoe, R., Ritoenour, T., Madugo, C., Malik, M. (2016). Shortening rate and Holocene surface rupture on the Riasi fault system in the Kashmir Himalaya: Active thrusting within the Northwest Himalayan orogenic wedge. *Geological Society of America Bulletin*, 128(7–8), 107–1094.

Gavillot, Y., Meigs, A.J., Sousa, F.J., Stockli, D., Yule, D., Malik, M. (2018). Late Cenozoic foreland-to-hinterland low-temperature exhumation history of the Kashmir Himalaya. *Tectonics*, 37, 3041–3068.

Ghani, H., Sobel, E.R., Zeilinger, G., Glodny, J., Zapata, S., Irum, I. (2020). Palaeozoic and Pliocene tectonic evolution of the salt range constrained by low-temperature thermochronology. *Terra Nova*, 33(3), 293–305.

Ghani, H., Sobel, E., Zeilinger, G., Glodny, J., Irum, I. (2021). Spatio-temporal structural evolution of the Kohat fold and thrust belt of Pakistan. *Journal of Structural Geology*, 145, 104310.

Grelaud, S., Sassi, W., de Lamotte, D.F., Jaswal, T., Roure, F. (2002). Kinematics of eastern salt range and south potwar basin (Pakistan): A new scenario. *Marine and Petroleum Geology*, 19(9), 1127–1139.

Hossack, J.R. (1979). The use of balanced cross-sections in the calculation of orogenic contraction: A review. *Journal of the Geological Society*, 136, 705–711.

Jouanne, F., Awan, A., Madji, A., Pêcher, A., Latif, M., Kausar, A., Mugnier, J.L., Khan, I., Khan, N.A. (2011). Postseismic deformation in Pakistan after the 8 October 2005 earthquake: Evidence of afterslip along a flat north of the Balakot-Bagh thrust. *Journal of Geophysical Research*, 116, B07401. doi:10.1029/2010JB007903.

Jouanne, F., Mugnier, J.L., Sapkota, S.N., Bascou, P., Pêcher, A. (2017). Estimation of coupling along the Main Himalayan Thrust in the central Himalaya. *Journal of Asian Earth Sciences*, 133, 62–71.

Jouanne, F., Munawar, N., Mugnier, J.L., Ahmed, A., Awan, A.A., Bascou, P., Vassallo, R. (2020). Seismic coupling quantified on inferred décollements beneath the western syntaxis of the Himalaya. *Tectonics*, 39(9), e2020TC006122.

Kaneda, H., Nakata, T., Tsutsumi, H., Kondo, H., Sugito, N., Awata, Y., Akhtar, S.S., Majid, A., Khattak, W., Awan, A.A. et al. (2008). Surface rupture of the 2005 Kashmir, Pakistan, earthquake and its active tectonic implications. *Bulletin of the Seismological Society of America*, 98, 521–557.

Khan, M.A., Bendick, R., Bhat, I.M., Bilham, R., Kakar, D.M., Khan, S.F., Lodi, S., Qazi, S., Singh, B., Szeliga, W., Wahab, A. (2008). Preliminary geodetic constraints on plate boundary deformation on the western edge of the Indian Plate from TriGGNnet. *Journal of Himalayan Earth Sciences*, 41, 71–87.

Kondo, H., Nakata, T., Akhtar, S., Wesnousky, S., Sugito, N., Kaneda, H., Tsutsumi, H., Khan, A., Khattak, W., Kausar, A. (2008). Long recurrence interval of faulting beyond the 2005 Kashmir earthquake around the northwestern margin of the Indo-Asian collision zone. *Geology*, 36, 731–734.

Leturmy, P., Mugnier, J.L., Baby, P., Coletta, B., Vinour, P. (2000). Interactions between thin skinned thrust tectonics and sedimentation: A view from numeric and analogue models. *Tectonophysics*, 320, 45–67.

Pathier, E., Fielding, E.J., Wright, T.J., Walker, R., Parsons, B.E., Hensley, S. (2006). Displacement field and slip distribution of the 2005 Kashmir earthquake from SAR imagery. *Geophysical Research Letters*, 33, L20310. doi:10.1029/2006GL027193.

Powers, P.M., Lillie, R.J., Yeats, R.S. (1998). Structure and shortening of the Kangra and Dehra Dun reentrants, Sub-Himalaya, India. *Geological Society of America Bulletin*, 110, 1010–1027.

Raiverman, V., Srivastava, A.K., Prasad, D.N. (1994). Structural style in northwestern Himalayan foothills. *Himalayan Geology*, 15, 263–280.

Satyabala, S.P., Yang, Z.H., Bilham, R. (2012). Stick–slip advance of the Kohat Plateau in Pakistan. *Nature Geoscience*, 5(2), 147–150.

Sreejith, K.M., Jasir, M.C.M., Ritesh, A., Rajawat, A.S. (2021). The 2019 September 24, Mw = 6, Mirpur earthquake, NW Himalaya: Geodetic evidence for shallow, near-horizontal décollement rupture of the Main Himalayan Thrust. *Tectonophysics*, 816, 229013.

Srinivasan, S. and Khar, B.M. (1996). Status of hydrocarbon exploration in northwest Himalaya and foredeep: Contribution to stratigraphy and structure. *Geological Survey of India Special Publication*, 21, 295–405.

Thakur, V.C., Jayangondaperumal, R., Malik, M.A. (2010). Redefining Medlicott-Wadia's main boundary fault from Jhelum to Yamuna, an active fault strand of the main boundary thrust in northwest Himalaya. *Tectonophysics*, 489, 29–42.

Thakur, V.C., Joshi, M., Jayangondaperumal, M. (2020). Active tectonics of Himalayan frontal fault zone in the Sub-Himalaya. In *Geodynamics of the Indian Plate*, Gupta, N. and Tandon, S.K. (eds). Springer, Cham.

Vassallo R., Mugnier, J.L., Vignon, V., Malik, M., Jayangondaperumal, R., Srivastava, P., Jouanne, F., Carcaillet, J. (2015). Quaternary deformation and seismic hazard in Northwestern Himalaya. *Earth and Planetary Science Letters*, 411, 241–252.

Vassallo, R., Mugnier, J.L., Jomard, H., Cortés-Aranda J., Malik, M., Jouanne, F., Buoncristiani, J.F. (2020). Recurrence of large paleo-earthquakes in Kashmir Himalaya seismic gap (Riasi area, India). *Journal of Asian Earth Sciences*, 201, 104505.

Vignon V., Mugnier, J.-L., Vassallo, R., Srivastava, P., Malik, M.A., Jayangondaperumal, R., Jouanne, F., Buoncristiani, J.-F., Carcaillet, J., Replumaz, A. et al. (2017). Sedimentation close to the active Medlicott Wadia Thrust (Western Himalaya): How to decipher tectonics and base level changes. *Geomorphology*, 284, 175–190.

Wang, K. and Fialko, Y. (2014). Space geodetic observations and models of postseismic deformation due to the 2005 M7.6 Kashmir (Pakistan) earthquake. *Journal of Geophysical Research: Solid Earth*, 119, 7306–7318. doi:10.1002/2014JB011122.

Yeats, R.S. and Lawrence, R.D. (1984). Tectonics of the Himalayan thrust belt in northern Pakistan. In *Marine Geology and Oceanography of Arabian Sea and Coastal Pakistan*, Haq, B.U. and Milliman, J.D. (eds). Von Nostrand Reinhold, New York.

8

The 2015 April 25 Gorkha Earthquake

Laurent BOLLINGER[1], Lok Bijaya ADHIKARI[2], Jérôme VERGNE[3], György HETÉNYI[4] and Shiba SUBEDI[4,5]

[1] *French Alternative Energies and Atomic Energy Commission, Bruyères-le-Châtel, France*

[2] *Department of Mines and Geology, Kathmandu, Nepal*

[3] *University of Strasbourg, France*

[4] *Institute of Earth Sciences, University of Lausanne, Switzerland*

[5] *Seismology at School in Nepal, Pokhara, Nepal*

8.1. Introduction

On April 25, 2015, a very large earthquake (magnitude 7.8) devastated central Nepal at the toe of the high Himalayan range (Figure 8.1). The seismic rupture initiated at depth on the Main Himalayan Thrust, it propagated eastwards for 50 seconds, accommodating a slip of several meters over an area of 140 km by 50 km. The earthquake severely affected the capital city Kathmandu. It left hundreds of thousands of locals homeless and killed nearly 10,000 people. This earthquake was followed by tens of thousands of aftershocks, including thousands of events felt by some of the 4 million inhabitants of the Kathmandu valley. The aftershocks developed around the region of the fault activated by the mainshock.

Himalaya, Dynamics of a Giant 3,
coordinated by Rodolphe CATTIN and Jean-Luc EPARD.

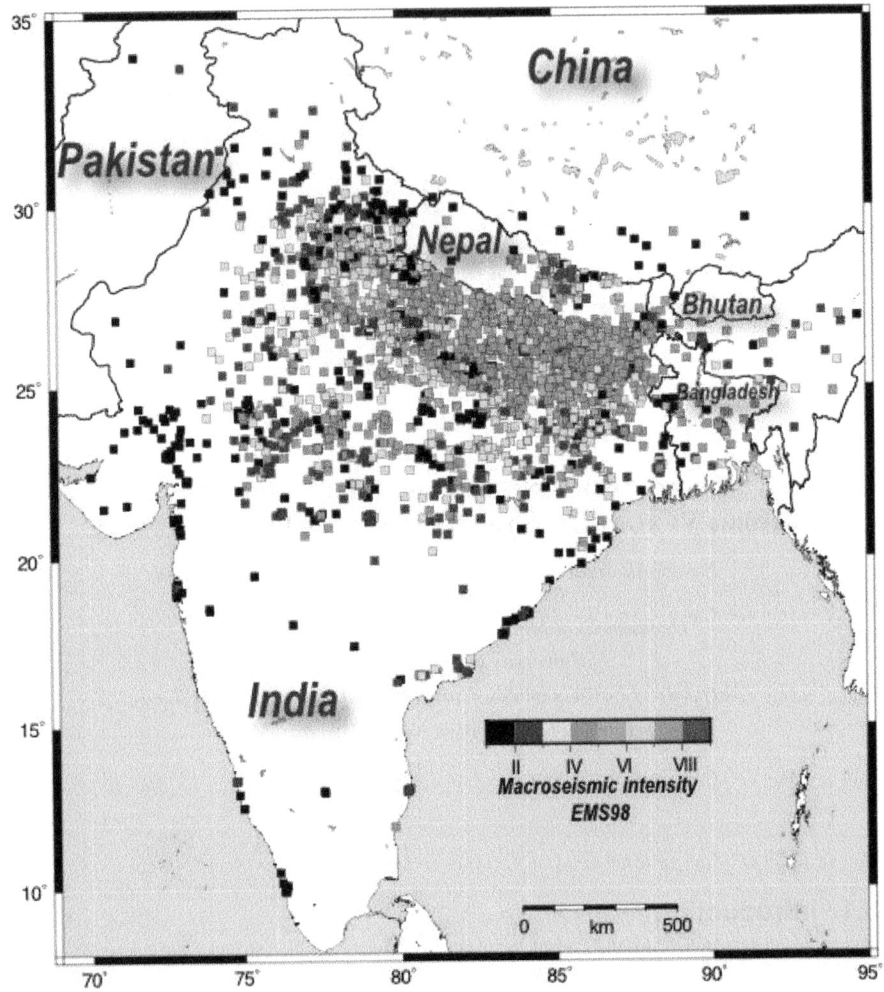

Figure 8.1. *Macroseismic intensities (European Macroseismic Scale EMS-98) for the April 25, 2015 Gorkha earthquake (intensities from Martin et al. 2015). For a color version of this figure, see www.iste.co.uk/cattin/himalaya3.zip*

Lateral variations of this seismicity coincide with changes in the deep geological structures along the fault plane (see Volume 1 – Chapter 4). These structures controlled the rupture propagation of the 2015 earthquake, and rupture only occurred on the northern half of the fault zone below central Nepal, leaving the southern half still locked and under high stresses.

The very strong seismic activity of the aftershocks decreased gradually with time, in accordance with post-seismic stress relaxation processes observed for other large intercontinental earthquakes. The seismicity of the region at the beginning of 2022 is still four times higher than it was before the 2015 earthquake. The analysis of the spatio-temporal variations of the aftershocks makes it possible to estimate that it probably will take another five years for the seismicity to return to its pre-earthquake rate unless a new major earthquake ruptures the fault segments that have been locked for hundreds of years to the south and west of the 2015 rupture area. In the following, we present how all this information is obtained from studies including seismology, geodesy and numerical modeling.

8.2. The mainshock and its effects on the ground

8.2.1. *Overview description of the mainshock*

On Saturday April 25, 2015, at 11 h 56 local time – a holiday at noon – a strong earthquake occurred in central Nepal. The mainshock was widely felt by over half of the Indian sub-continent (Figure 8.1). The very large area where the event was felt testifies for the large magnitude of the event. After a few minutes, the Nepalese seismological institutes confirm a magnitude larger than 7.5. Ultimately, the magnitude is estimated at between 7.8 and 7.9, according to the seismic moment released during the dozens of seconds of the rupture propagation on the fault.

The mesoseismal area – the region affected by the most important damage – develops over a length of about 100 km along the strike of the Himalaya, at the toe of the high mountain range's southern slopes. These slopes were impacted by thousands of landslides (Roback et al. 2018), among which the terrible Langtang valley landslide completely washed out Langtang village, killing more than half of its inhabitants (Kargel et al. 2016, see also Volume 3 – Chapter 4 for landslides-earthquakes relationships).

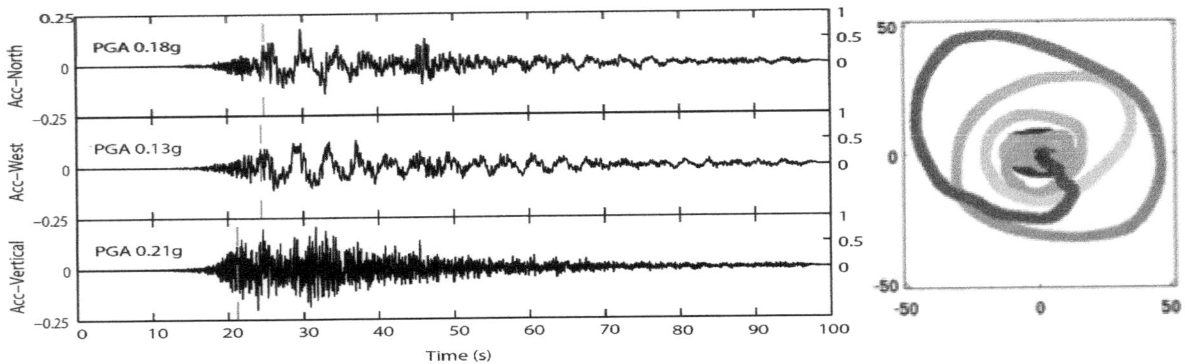

Figure 8.2. *Ground motion following the 2015 Gorkha earthquake. (Left) North, West and Down components of the acceleration time series recorded for the Gorkha earthquake (expressed in g) in Kathmandu at the accelerometer of the Department of Mines and Geology (modified from Bhattarai et al. (2015)). (Right) From red to blue: horizontal displacement in cm during the 70s that followed the onset of the P-wave. For a color version of this figure, see www.iste.co.uk/cattin/himalaya3.zip*

Kathmandu lies 30 km south of the range, in the middle of the mesoseismal zone. The first seconds of the shaking were documented there by hundreds of smartphones and surveillance camera videos which captured spectacular motion of crowds gathered outside, swaying back and forth, sometime circularly.

This strong ground motion was also quantified thanks to a few geodetic and accelerometric instruments available in the valley (Figure 8.2) and installed at rock sites. The stations located above the thick lacustrine sediments of the Kathmandu basin recorded peak ground accelerations lower than expected but a longer duration of the strong motion than at rock sites (see Volume 3 – Chapter 6). This observation was later interpreted and modeled as associated with the trapping of the seismic waves in the sedimentary basin due to softer sediments between and above stronger rocks (mountains).

The macroseismic intensities estimated in the valley were lower than expected (Martin et al. 2015; Hough et al. 2016; Rajaure et al. 2017). However, despite these specific observations in the capital city, the toll of the earthquake was heavy. The damages related to the earthquake shaking and its effects were impressive, with 300,000 houses heavily affected. A large number of victims were found under collapsed buildings and landslides, particularly in the villages with vulnerable masonry. The shaking and damages resulted in nearly 10,000 persons being killed, according to the official estimates, and thousands being displaced. Hundreds of thousands remained homeless during weeks, living outside under tents above the rupture fault patch, in a region impacted by frequent aftershocks. Nevertheless, these numbers remain below those that could have been the case in an earthquake of higher magnitude, higher frequency shaking or more centered on Kathmandu.

8.2.2. *Ground motion and deformation*

As previously mentioned, the ground motion dynamics were recorded at seismic and high-rate GPS stations. These instruments complemented locally the measurements of the static ground deformations determined by the acquisition of interferometrisable couples of Synthetic Aperture Radar images (known as the InSAR method).

The GPS stations installed in the mesoseismal area were shifted towards the south, the instruments along the northern side of the region subsided, while most instruments installed in the vicinity of the Kathmandu valley were uplifted, sometimes by dozens of centimeters (Figure 8.3).

InSAR images revealed that the Gosainkund–Langtang range subsidence exceeded locally a meter, while the Kathmandu valley and Shivapuri Lekh was uplifted significantly, somewhere by more than a meter (Figure 8.3; e.g. Grandin et al. 2015; Kobayashi et al. 2015; Lindsey et al. 2015).

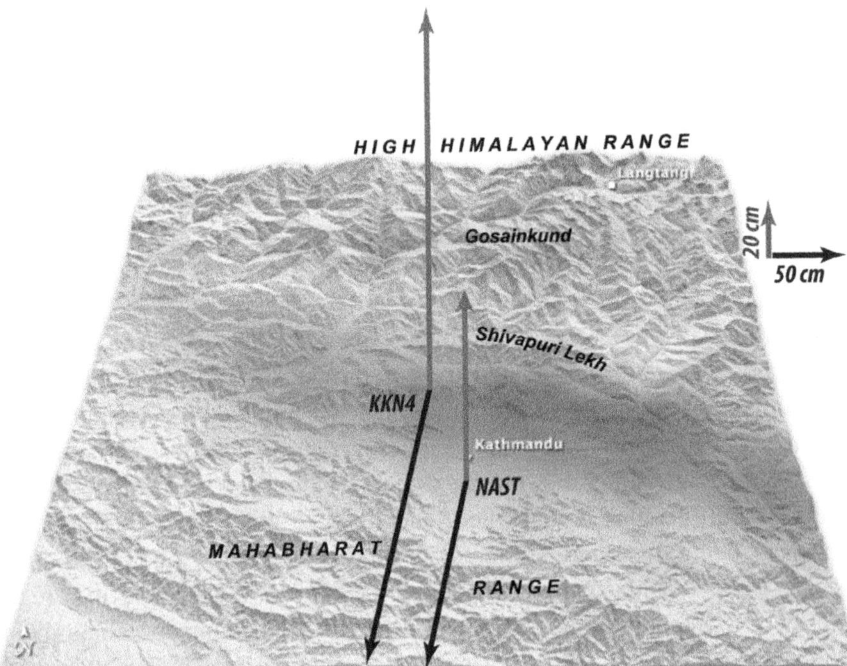

Figure 8.3. *Coseismic displacement field seen from remote sensing (here European Space Agency Sentinel-1 InSAR block model of the ground displacement) (source: ESA – and GPS horizontal and vertical displacement – in black and red (Galetzka et al. 2015)). KKN4 and NAST are two continuous GPS stations operated near Kathmandu. For a color version of this figure, see www.iste.co.uk/cattin/himalaya3.zip*

8.3. Investigating the seismic source at depth

The Gorkha earthquake as well as its largest aftershocks were recorded at hundreds of distant seismometers, complementing the InSAR measurements of the deformation and the local seismological and GPS records of the ground motion.

Seismological analysis of the mainshock (using a technique called "back-projection") was performed at several regional arrays of stations. This technique provided successive images of coherent sources of seismic waves which nucleated during the fault rupture at depth.

Several studies were obtained by kinematic inversions of these teleseismic waves jointly or independently of the local dataset (e.g. Avouac et al. 2015; Grandin et al. 2015), and point towards a robust, first-order knowledge of the seismic slip released at depth.

The kinematic histories which resulted from these works demonstrate that the rupture propagated mainly unilaterally towards the East, from a focal/nucleation point best constrained by the national seismological center – and often fixed by the authors in their inversions. This nucleation occurred at a depth of 13–15 km, near the down-dip end of the locked Main Himalayan Thrust fault, beneath the toe of the high Himalaya and the village of Barpak. The onset of the rupture was followed by a low and relatively slow slip start. The amount of slip released increased significantly, from 1 to 5 meters within the first 50 km. The coseismic slip reaches a peak, exceeding 5 meters North of Kathmandu over a 50 by 20-km large fault patch. The highest resolution inversions demonstrate the presence of two slip maxima, exceeding 6 meters, that ruptured after 20–30s and 30–40s (Grandin et al. 2015, Figure 8.4). The total duration of the mainshock appears to be close to 50 seconds during which the released seismic moment reached $\sim 7.5 \times 10^{20}$ N.m corresponding to a moment magnitude Mw of approximately 7.85.

The coseismic slip at depth on the MHT decreased abruptly at about half the distance towards the frontal thrust, leaving approximately the southern half of the fault unruptured. This relatively abrupt ending of the coseismic slip beneath Kathmandu was later attributed by some authors to the presence of a 2–3 km-high midcrustal ramp, consistent with geologically balanced cross-sections (Hubbard et al. 2016). Overall, several complementary observations along the strike demonstrated that the rupture associated with the mainshock was controlled structurally.

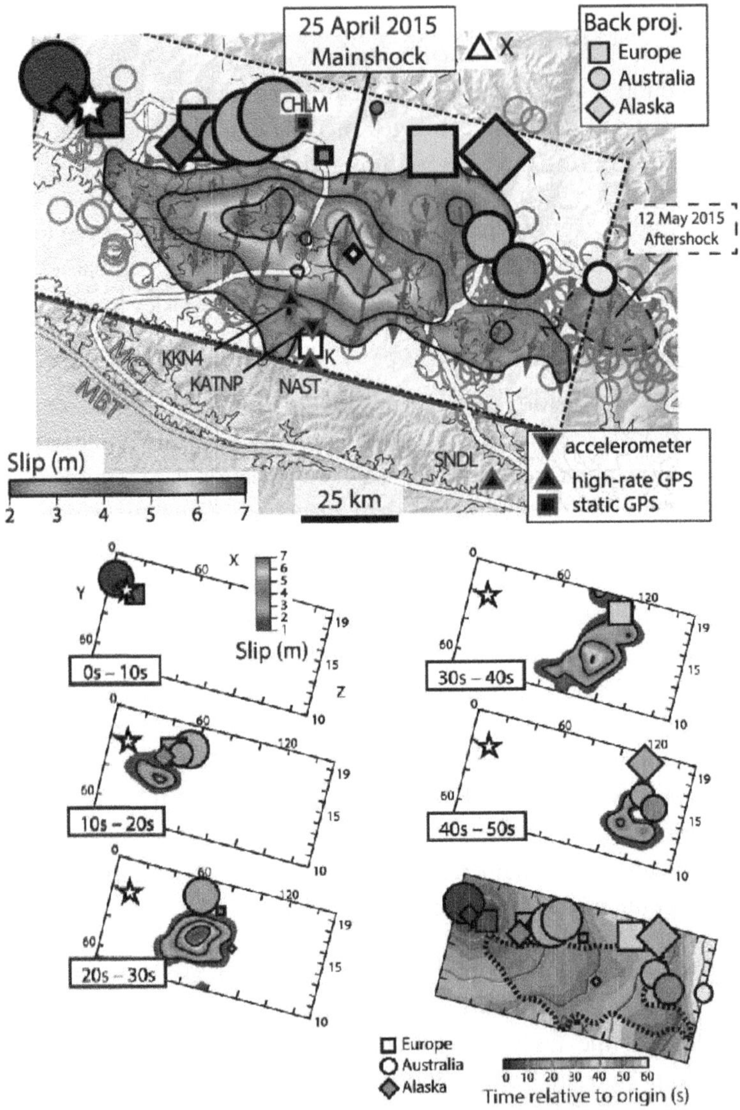

Figure 8.4. *Slip distribution of the Gorkha earthquake, with 2 meters slip contours and snapshots of the slip at various times within 50 seconds from the earthquake nucleation (modified from Grandin et al. 2015). For a color version of this figure, see www.iste.co.uk/cattin/himalaya3.zip*

8.4. Aftershock activity and post-seismic relaxation

8.4.1. *Early aftershocks and the occurrence of the Kodari earthquake*

The mainshock was immediately followed by aftershocks that occurred around the whole stretch of the Main Himalayan Thrust fault segments which ruptured at depth (Adhikari et al. 2015, Figure 8.5). The aftershocks happened at close vicinity to the mid-crustal shear zone in cross-section (Duputel et al. 2016).

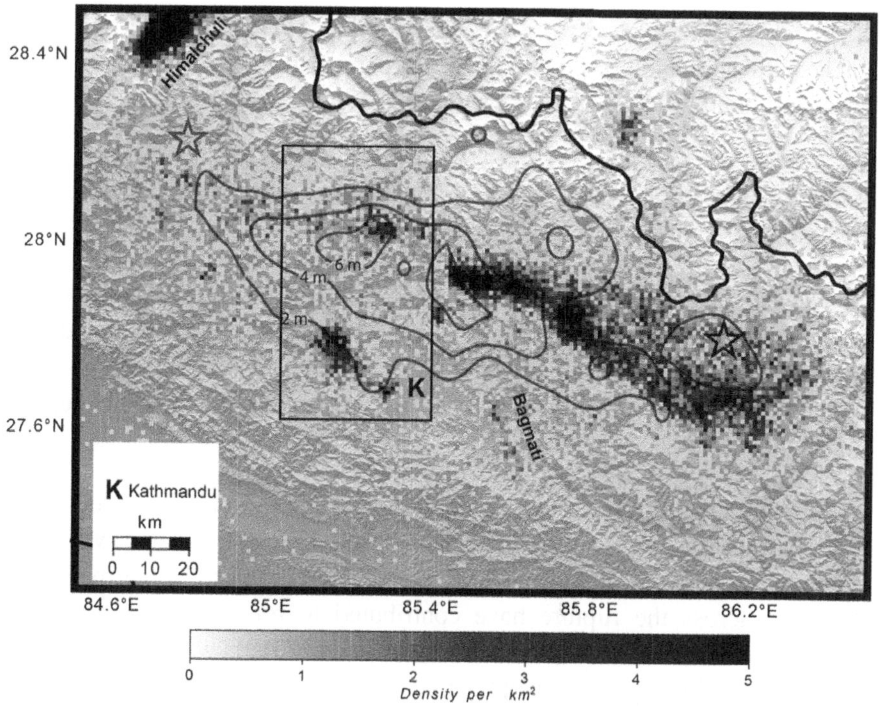

Figure 8.5. *Density map of five years of aftershocks from (Adhikari et al. 2022). Red star: epicenter of the April 25, 2015 Gorkha earthquake (M7.9). Black star: epicenter of the May 12, 2015 Kodari earthquake (M7.3). The rectangle localizes the frame of Figure 8.6. For a color version of this figure, see www.iste.co.uk/cattin/himalaya3.zip*

These early aftershocks were followed on May 12 by a large earthquake of magnitude M 7.3, the Kodari earthquake, that ruptured a more than 20 km long fault patch at the eastern end of the region affected by the Gorkha earthquake rupture. This rupture zone completes that of the April 25 rupture zone to a rectangular shape, and therefore, the May 12 event is considered as a triggered earthquakes rather than a true aftershock. The subsequent aftershock activity in the region was then significantly revived. The national network detected 553 earthquakes of local magnitude greater than 4.0 within the first 45 days after the mainshock, (Adhikari et al. 2015), a number reaching 947 within the first five years. More than 31,000 smaller aftershocks were located during the same time period (Adhikari et al. 2021). The Gorkha aftershock sequence still defines a $\sim$140-km long E-SE trending structure, principally located at a depth beneath the southern slope of the high mountain range.

In addition, we observe a second seismicity belt located in the south, under the Kathmandu basin and in the northern part of the Mahabharat range (Adhikari et al. 2015, Figures 8.5 and 8.6).

8.4.2. *Seismicity monitored by denser, dedicated temporary networks*

The national network was robust enough to resist to the mainshock and kept monitoring the entire seismic sequence. However, the spatial density of stations and the instruments available at that time (mostly short-period vertical-component seismometers) were not optimal to constrain the depth of the earthquakes.

More than 100 temporary seismological stations were installed in the aftermath of the earthquake, complementing the seismic observations at the national network (Figure 8.7). Some of these stations which have been deployed across the rupture have contributed to reflection imaging of the geological structures at depth (Kurashimo et al. 2019), while others helped further constrain the geometry of the mid-crustal aftershocks clusters (Bai et al. 2019; Mendoza et al. 2019; Yamada et al. 2020). The spatial variations of the aftershocks along strike further revealed a 3D structural and geological control of the seismicity (e.g. Letort et al. 2016; Baillard et al. 2017; Wang et al. 2017). Stations deployed elsewhere along strike the Main Himalayan Thrust helped reveal the shallow seismicity and its control by mid-crustal structures, a result that seems ubiquitous (Hoste-Colomer et al. 2018; Laporte et al. 2021).

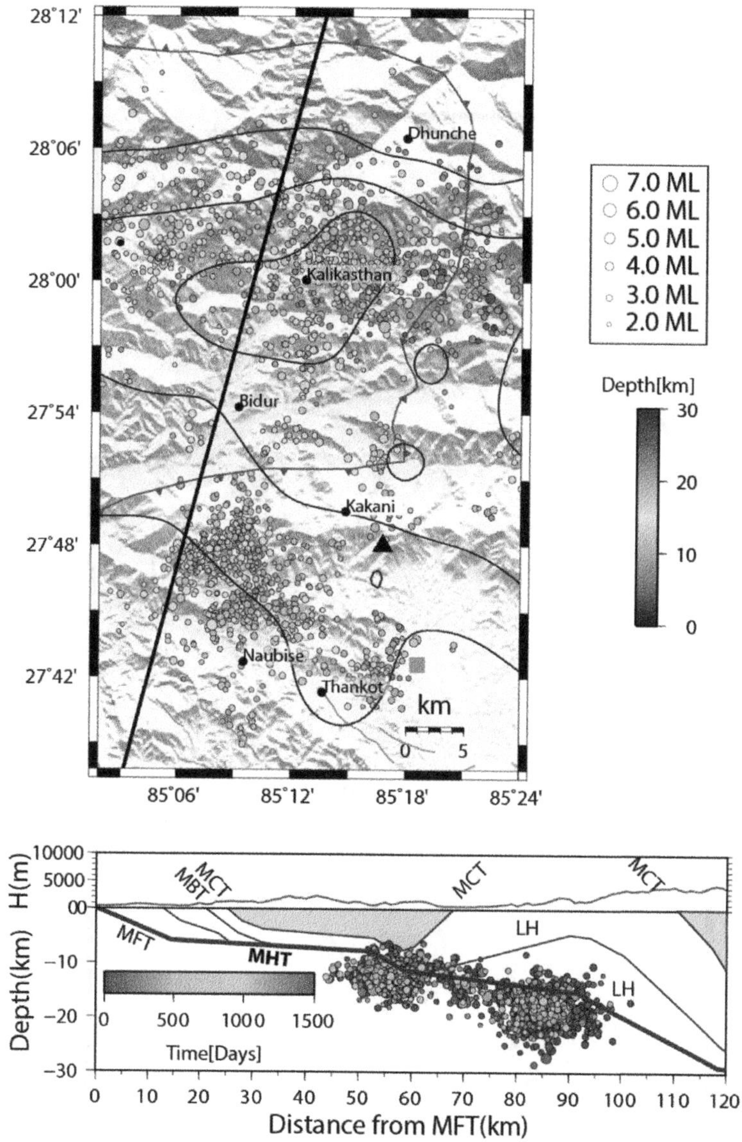

Figure 8.6. *Seismicity map and cross-section through the aftershocks located in the Kathmandu–Langtang region (modified from Adhikari et al. (2022)). Blue contours materialize the 2-meter isocontours of the coseismic slip. Projection of the seismicity on the balanced cross-section of Hubbard et al. (2016). For a color version of this figure, see www.iste.co.uk/cattin/himalaya3.zip*

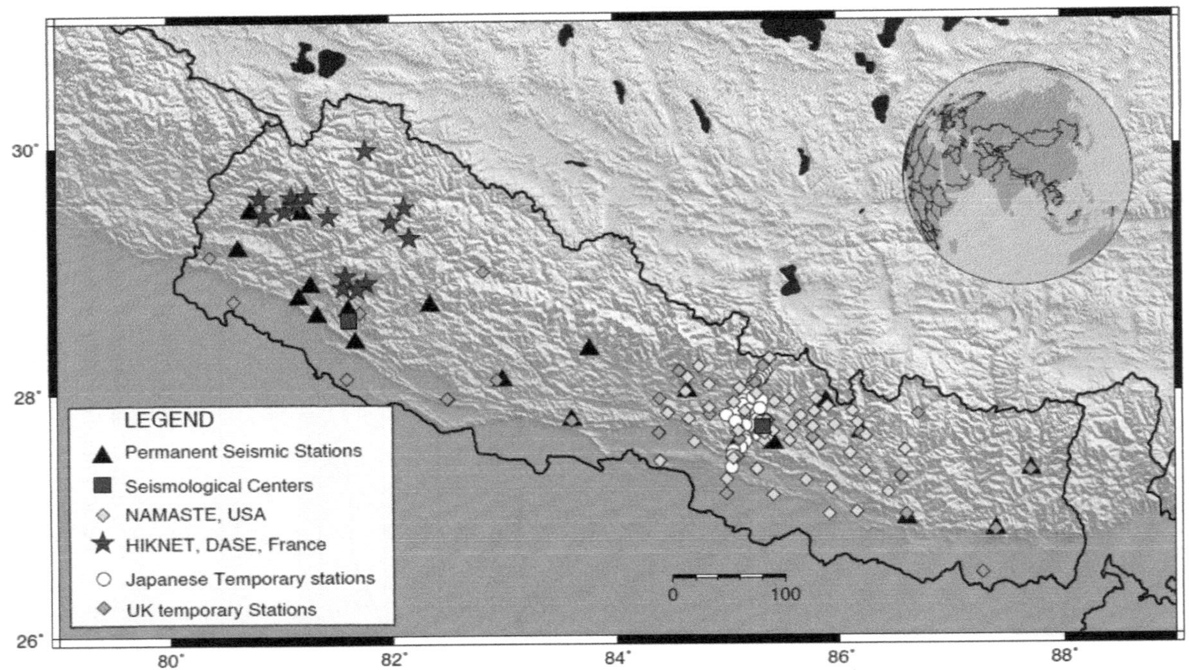

Figure 8.7. *Seismological stations deployed during the two years that followed the 2015 Gorkha earthquake mainshock, complementing the permanent seismic stations (black triangles). For a color version of this figure, see www.iste.co.uk/cattin/himalaya3.zip*

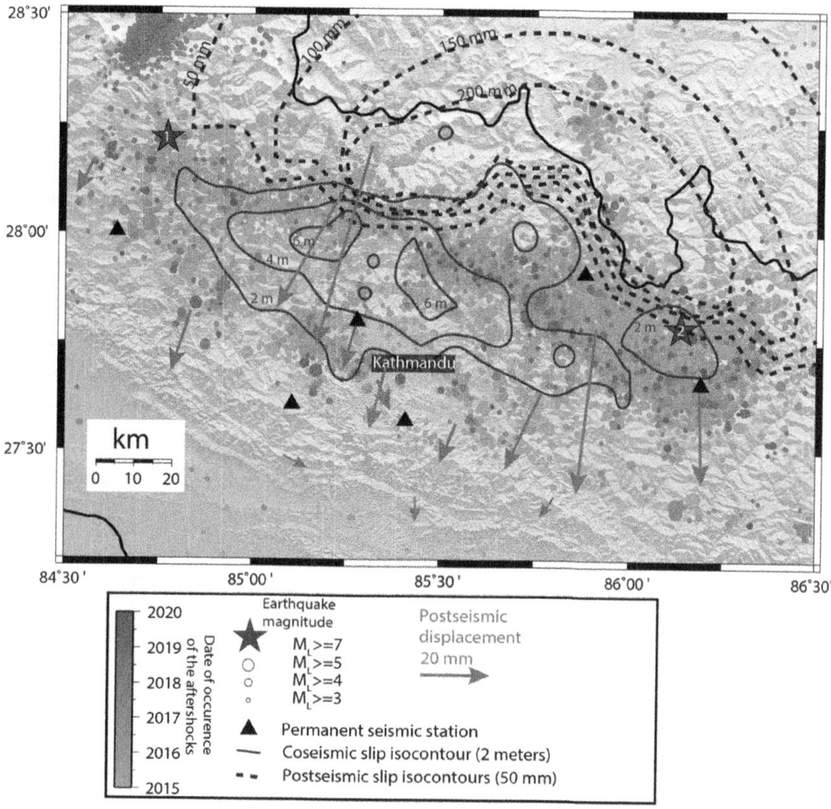

Figure 8.8. *Aftershocks located by the national network after the Gorkha earthquake (modified from Adhikari et al. 2022), and post-seismic deformation occurring north of that area (contours (Zhao et al. 2017) and arrows (Gualandi et al. 2017)). For a color version of this figure, see www.iste.co.uk/cattin/himalaya3.zip*

8.4.3. *Post-seismic relaxation monitored by geodetic and seismological networks*

The post-seismic deformation in the area of the rupture was monitored by a permanent, continuously operated GPS network, complemented by an array of temporary stations which helped monitoring for months following the earthquake. The spatial structure of the overall displacement field revealed that the post-seismic slip occurs, beyond the fault-locked patches, on the deeper basal flat of the MHT, and it occurs mostly in an aseismic way (Figure 8.8; e.g. Gualandi et al. 2017; Zhao et al. 2017; Jouanne et al. 2019; Liu-Zeng et al. 2020). This deep, aseismic relaxation is still temporally associated with a persistent seismic activity along the coseismic rupture (Figure 8.8).

The upper part of the locked fault zone, which remained unruptured during the mainshock, remained fully locked during the seven years that followed the earthquake demonstrating that the region is still at risk of large ruptures despite the recent seismic crisis.

After seven years of aftershock activity, the annual rate of earthquakes with magnitudes above ML 2.5 (the detection threshold corresponding to the present magnitude of completeness of the Nepalese national seismicity catalogue) is still four times higher than it was before the 2015 earthquake (Figure 8.9). This is still consistent with the conclusion of early works on post-seismic relaxation constrained by geodesy (e.g. Gualandi et al. 2017), which estimated that the relaxation phase will take more than a decade.

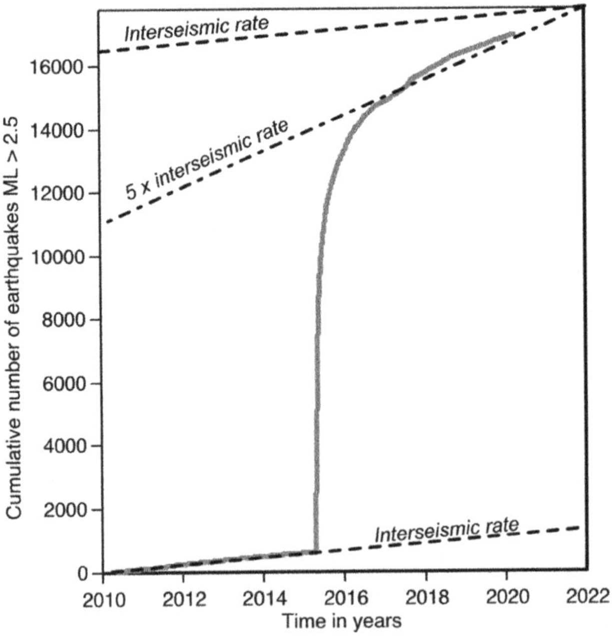

Figure 8.9. *Cumulative number of earthquakes in the rupture area of the 2015 Gorkha earthquake. The seismicity catalogue was cut at ML 2.5, a magnitude threshold under which the catalogue remains incomplete. The rate of seismicity is still more than four times larger than before the 2015 earthquake (modified from Adhikari et al. (2022))*

In summary, the post-seismic phase is characterized mainly by deep afterslip, but almost no transfer of stresses or accommodation of aseismic slip on the upper, locked segment of the Main Himalayan Thrust.

8.5. A more earthquake-informed and earthquake-resilient local community in the aftermath of the earthquake

The scientific knowledge of the mainshock and aftershocks was documented in hundreds of peer-reviewed publications. However, this and earlier scientific knowledge on Himalayan geology and seismotectonics took and still take a long time to be disseminated to the public.

The demand of the public for early earthquake information was tremendous, particularly in the aftermath of the mainshock – the first weeks following the event being accompanied by a high rate of felt earthquakes in the Kathmandu valley. At that time, the institution in charge of seismic alerts had to deliver the basic information about the time, location and magnitude of the largest aftershocks on a phone server, making it available to the public. This period of anxiety for the local inhabitants corresponded with the development of the use of the LastQuake app, massively downloaded in Nepal (Bossu et al. 2015). The scientific information were also percolating to the public after the technical workshop organized by the research institutes in Nepal, and relayed by the media. However, these efforts mainly reached those who were actively seeking information, and had access to communication pathways and direct sources. Relatively soon, especially in the countryside, other, daily challenges became the primary focus of the local population.

It is in this context that a new effort has initiated in Nepal, with the goal of preparing the next generation of Nepali for the next generation of earthquakes (Figure 8.10). The best way to reach the corresponding group of population are the schools: today's students are tomorrow's parents, house builders and decision-makers. Therefore, a bottom-up initiative started the Seismology at School in Nepal program (Subedi et al. 2020). The departure point is simple: train the teachers so that they can teach earthquake topics in the schools, and bring low-cost seismometers at schools to create attraction and to use them in the classroom. The area where the program started was chose near the epicenter of the 2015 Gorkha earthquake. The implementation worked out thanks to the excellent local connections between, teachers, schools, students and the local administration. The project has not only grown in size, reaching 40 schools in 2022, but has also diversified beyond the classroom material. Flyers and tickers have been distributed, a card game on earthquakes have been developed, the local traditions and Hindu religious concepts on earthquakes have been discussed, and the now famous Earthquake Awareness Song has been released (https://www.youtube.com/watch?v=ymE-lrAK0TI). The feedback from the locals is overwhelming, and it is hoped that the good example is spread further across the country. The authors of this book have given their consent to transfer all royalties to Seismology at School in Nepal.

Figure 8.10. *Left: a Nepali girl who participates in the Seismology at School in Nepal program (credit: Shiba Subedi). Right: a gentle reminder sticker about earthquakes, distributed in over 6,000 examples to school students in Nepal. For a color version of this figure, see www.iste.co.uk/cattin/himalaya3.zip*

8.6. Conclusion

The 2015 Gorkha earthquake is so far the instrumentally best-observed among the large Himalayan earthquakes (e.g. Adhikari 2021). The scientific knowledge gathered from this earthquake has already allowed scientists to reconsider some historical earthquake testimonies and the previously associated parameters. Among those, the 1833AD earthquake that presented some analogies to the 2015 event, and is now recognized as a similar earthquake (Martin et al. 2015).

The documentation of the structural and geological control on the mainshock rupture, on the spatial variations of the aftershocks and on interseismic earthquakes that happened in the decades prior the earthquake (Adhikari et al. 2022, and Volume 3 – Chapter 6) helped to revise the structural segmentation of the thrust fault, and its impact on seismic hazard assessment (Stevens et al. 2018; Michel et al. 2021). The improved knowledge of the aftershock sequence, post-mainshock seismic and aseismic straining already helped to determine more accurately the seismic moment budget along the fault system.

The first years of monitoring of the post-seismic relaxation confirm that the seismic moment accumulated during centuries of stress build-up at mid-crustal depth along the Main Himalayan Thrust is ultimately released by large and great earthquakes only. More regionally, despite its large magnitude, around 7.9, the rupture of the Gorkha earthquake has not propagated beyond more than half the N-S width of the locked fault zone. In adjacent zone, the fault patches remain unruptured since at least 1255 or 1344 AD to the South and 1344 or 1505 AD to the West, and 1505 under Far Western Nepal (see Volume 3 – Chapter 5). A large seismogenic potential is therefore already available in the region for nucleating large and great devastating earthquakes. It is therefore highly desired that research and monitoring effort focus on the Himalaya not only in the years after devastating events but continuously, and that educational and outreach efforts scale up to the whole population with the active help of every level of governance.

8.7. References

Adhikari, L.B. (2021). Seismicity associated with the April 25, 2015 Gorkha earthquake in Nepal: Probing the Himalayan Seismic Cycle.

Adhikari, L.B., Gautam, U.P., Koirala, B.P., Bhattarai, M., Kandel, T.P., Gupta, R.M., Timsina, C., Maharjan, N., Maharjan, K., Dahal, T. et al. (2015). The aftershock sequence of the April 25th 2015 Gorkha-Nepal earthquake. *Geophysical Journal International*, 203, 2119–2124.

Adhikari, L.B., Bollinger, L., Vergne, J., Lambotte, S., Chanard, K., Laporte, M., Li, L., Koirala, B.P., Bhattarai, M., Timsina, M. et al. (2021). Orogenic collapse and stress adjustments revealed by an intense seismic swarm following the 2015 Gorkha earthquake in Nepal. PhD thesis, University of Paris, Institut de Physique du Globe de Paris, France, *Frontiers in Earth Science*, 9(659937), 1–13.

Adhikari, L.B., Laporte, M., Bollinger, L., Vergne, J., Lambotte, S., Koirala, B.P., Bhattarai, M., Timsina, C., Gupta, R.M., Wendling-Vazquez, N. et al. (2022, in press). Seismically active structures of the main Himalayan thrust revealed before, during and after the 2015 Mw 7.9 Gorkha earthquake in Nepal. *Geophysical Journal International*, ggac281. https://doi.org/10.1093/gji/ggac281.

Avouac, J.-P., Meng, L., Wei, S., Wang, T., Ampuero, J. (2015). Lower edge of locked main Himalayan thrust unzipped by the 2015 Gorkha earthquake. *Nature Geoscience*, 8(9), 708–711.

Bai, L., Klemperer, S.L., Mori, J., Karplus, M.S., Ding, L., Liu, H., Li, G., Song, B., Dhakal, S. (2019). Lateral variation of the main Himalayan thrust controls the rupture length of the 2015 Gorkha earthquake in Nepal. *Science Advances*, 5(6), eaav0723.

Baillard, C., Lyon-Caen, H., Bollinger, L., Rietbrock, A., Letort, J., Adhikari, L.B. (2017). Automatic analysis of the Gorkha earthquake aftershock sequence: Evidences of structurally segmented seismicity. *Geophysical Journal International*, 209(2), 1111–1125.

Bhattarai, M., Adhikari, L.B., Gautam, U.P., Laurendeau, A., Labonne, C., Hoste-Colomer, R., Sèbe, O., Hernandez, B. (2015). Overview of the large 25 April 2015 Gorkha, Nepal, earthquake from accelerometric perspectives. *Seismological Research Letters*, 86(6), 1540–1548.

Bossu, R., Laurin, M., Mazet-Roux, G., Roussel, F., Steed, R. (2015). The importance of smartphones as public earthquake-information tools and tools for the rapid engagement with eyewitnesses: A case study of the 2015 Nepal earthquake sequence. *Seismological Research Letters*, 86(6), 1587–1592.

Duputel, Z., Vergne, J., Rivera, L., Wittlinger, G., Farra, V., Hetényi, G. (2016). The 2015 Gorkha earthquake: A large event illuminating the main Himalayan thrust fault. *Geophysical Research Letters*, 43, 2517–2525.

Galetzka, J., Melgar, D., Genrich, J., Geng, J., Owen, S., Lindsey, E., Xu, X., Bock, Y., Avouac, J.-P., Adhikari, L. et al. (2015). Slip pulse and resonance of Kathmandu basin during the 2015 Mw 7.8 Gorkha earthquake, Nepal imaged with geodesy. *Science*, 3490(6252), 1091–1095.

Grandin, R., Vallée, M., Satriano, C., Lacassin, R., Klinger, Y., Simoes, M., Bollinger, L. (2015). Rupture process of the Mw = 7.9 2015 Gorkha earthquake (Nepal): Insights into Himalayan megathrust segmentation. *Geophysical Research Letters*, 42(20), 8373–8382.

Gualandi, A., Avouac, J.-P., Galetzka, J., Genrich, J.F., Blewitt, G., Adhikari, L.B., Koirala, B.P., Gupta, R., Upreti, B.N., Pratt-Sitaula, B. et al. (2017). Pre- and post-seismic deformation related to the 2015, Mw7.8 Gorkha earthquake, Nepal. *Tectonophysics*, 714–715, 90–106.

Hoste-Colomer, R., Bollinger, L., Lyon-Caen, H., Adhikari, L.B., Baillard, C., Benoit, A., Bhattarai, M., Gupta, R.M., Jacques, E., Kandel, T. et al. (2018). Lateral variations of the midcrustal seismicity in western Nepal: Seismotectonic implications. *Earth and Planetary Science Letters*, 504, 115–125.

Hough, S.E., Martin, S.S., Gahalaut, V., Joshi, A., Landes, M., Bossu, R. (2016). A comparison of observed and predicted ground motions from the 2015 MW7. 8 Gorkha, Nepal, earthquake. *Natural Hazards*, 84(3), 1661–1684.

Hubbard, J., Almeida, R., Foster, A., Sapkota, S.N., Bürgi, P., Tapponnier, P. (2016). Structural segmentation controlled the 2015 M w 7.8 Gorkha earthquake rupture in Nepal. *Geology*, 44(8), 639–642.

Jouanne, F., Gajurel, A.P., Mugnier, J.L., Bollinger, L., Adhikari, L.B., Koirala, B., Cotte, N., Bhattari, R., Pecher, A., Bascou, P., Huyghe, P. (2019). Postseismic deformation following the April 25, 2015 Gorkha earthquake (Nepal): Afterslip versus viscous relaxation. *Journal of Asian Earth Sciences*, 176, 105–119.

Kargel, J.S., Leonard, G.J., Shugar, D.H., Haritashya, U.K., Bevington, A., Fielding, E.J., Fujita, K., Geertsema, M., Miles, E.S., Steiner, J. et al. (2016). Geomorphic and geologic controls of geohazards induced by Nepal's 2015 Gorkha earthquake. *Science*, 351(6269), aac8353.

Kobayashi, T., Morishita, Y., Yaray, H. (2015). Detailed crustal deformation and fault rupture of the 2015 Gorkha earthquake, Nepal, revealed from ScanSAR-based interferograms of ALOS-2. *Earth Planets Space*, 670(201), 1–13.

Kurashimo, E., Sato, H., Sakai, S.I., Hirata, N., Gajurel, A.P., Adhikari, D.P., Subedi, K.P., Yagi, H., Upreti, B.N. (2019). The 2015 Gorkha earthquake: Earthquake reflection imaging of the source fault and connecting seismic structure with fault slip behavior. *Geophysical Research Letters*, 46(6), 3206–3215.

Laporte, M., Bollinger, L., Lyon-Caen, H., Hoste-Colomer, R., Duverger, C., Letort, J., Riesner, M., Koirala, B.P., Bhattarai, M., Kandel, T. et al. (2021). Seismicity in far western Nepal reveals flats and ramps along the Main Himalayan Thrust. *Geophysical Journal International*, 226(3), 1747–1763.

Letort, J., Bollinger, L., Lyon-Caen, H., Guilhem, A., Cano, Y., Baillard, C., Adhikari, L.B. (2016). Teleseismic depth estimation of the 2015 Gorkha – Nepal aftershocks. *Geophysical Journal International*, 207(3), 1584–1595.

Lindsey, E.O., Natsuaki, R., Xu, X., Shimada, M., Hashimoto, M., Melgar, D., Sandwell, D.T. (2015). Line-of-sight displacement from ALOS-2 interferometry: Mw 7.8 Gorkha earthquake and Mw 7.3 aftershock. *Geophysical Research Letters*, 42, 6655–6661.

Liu-Zeng, J., Zhang, Z., Rollins, C., Gualandi, A., Avouac, J.-P., Shi, H., Wang, P., Chen, W., Zhang, R., Zhang, P. et al. (2020). Postseismic deformation following the 2015 Mw7. 8 Gorkha (Nepal) earthquake: New GPS data, kinematic and dynamic models, and the roles of afterslip and viscoelastic relaxation. *Journal of Geophysical Research: Solid Earth*, 125(9), e2020JB019852.

Martin, S., Hough, S.E., Hung, C. (2015). Ground motions from the 2015 Mw 7.8 Gorkha, Nepal, earthquake constrained by a detailed assessment of macroseismic data. *Seismological Research Letters*, 86, 1524–1532. doi:10.1785/0220150138.

Mendoza, M.M., Ghosh, A., Karplus, M.S., Klemperer, S.L., Sapkota, S.N., Adhikari, L.B., Velasco, A. (2019). Duplex in the Main Himalayan Thrust illuminated by aftershocks of the 2015 Mw 7.8 Gorkha earthquake. *Nature Geoscience*, 12(12), 1018–1022.

Michel, S., Jolivet, R., Rollins, C., Jara, J., Dal Zilio, L. (2021). Seismogenic potential of the main Himalayan thrust constrained by coupling segmentation and earthquake scaling. *Geophysical Research Letters*, 48(13), e2021GL093106.

Rajaure, S., Asimaki, D., Thompson, E.M., Hough, S., Martin, S., Ampuero, J.P., Dhital, M.R., Inbal, A., Takai, N., Shigefuji, M. et al. (2017). Characterizing the Kathmandu valley sediment response through strong motion recordings of the 2015 Gorkha earthquake sequence. *Tectonophysics*, 714, 146–157.

Roback, K., Clark, M.K., West, A.J., Zekkos, D., Li, G., Gallen, S.F., Chamlagain, D., Godt, J.W. (2018). The size, distribution, and mobility of landslides caused by the 2015 Mw7.8 Gorkha earthquake, Nepal. *Geomorphology*, 301, 121–138.

Stevens, V.L., Shrestha, S.N., Maharjan, D.K. (2018). Probabilistic seismic hazard assessment of Nepal. *Bulletin of the Seismological Society of America*, 108(6), 3488–3510.

Subedi, S., Hetényi, G., Denton, P., Sauron, A. (2020). Seismology at school in Nepal: A program for educational and citizen seismology through a low-cost seismic network. *Frontiers in Earth Science*, 73, 1–19.

Wang, X., Wei, S., Wu, W. (2017). Double-ramp on the main Himalayan thrust revealed by broadband waveform modeling of the 2015 Gorkha earthquake sequence. *Earth and Planetary Science Letters*, 473, 83–93.

Yamada, M., Kandel, T., Tamaribuchi, K., Ghosh, A. (2020). 3D fault structure inferred from a refined aftershock catalog for the 2015 Gorkha earthquake in Nepal. *Bulletin of the Seismological Society of America*, 110(1), 26–37.

Zhao, B., Bürgmann, R., Wang, D., Tan, K., Du, R., Zhang, R. (2017). Dominant controls of downdip afterslip and viscous relaxation on the postseismic displacements following the Mw7.9 Gorkha, Nepal, earthquake. *Journal of Geophysical Research: Solid Earth*, 122(10), 8376–8401.

9

Crustal Fluids in the Nepal Himalaya and Sensitivity to the Earthquake Cycle

Frédéric GIRAULT[1], Christian FRANCE-LANORD[2], Lok Bijaya ADHIKARI[3], Bishal Nath UPRETI[4], Kabi Raj PAUDYAL[5], Ananta Prasad GAJUREL[5], Pierre AGRINIER[1], Rémi LOSNO[1], Chiara GROPPO[6,7], Franco ROLFO[6,7], Sandeep THAPA[1], Shashi TAMANG[6,1,5] and Frédéric PERRIER[1]

[1] *Paris Cité University, France*
[2] *University of Nancy, Vandoeuvre-lès-Nancy, France*
[3] *Department of Mines and Geology, Kathmandu, Nepal*
[4] *Nepal Academy of Science and Technology, Kathmandu, Nepal*
[5] *Tribhuvan University, Kathmandu, Nepal*
[6] *Department of Earth Sciences, University of Turin, Italy*
[7] *Institute of Geoscience and Geohazard, Turin, Italy*

9.1. Introduction

Based on the unprecedented dataset on hydrothermal systems presented in detail in Volume 1 – Chapter 8, we can explore the characteristics and propose interpretations on thermal springs and gaseous emissions in the Nepal Himalaya. Datasets of thermal springs and gaseous emissions are summarized in the supplementary material www.iste.co.uk/cattin/himalaya3.zip.

Himalaya, Dynamics of a Giant 3,
coordinated by Rodolphe CATTIN and Jean-Luc EPARD.

In this chapter, we present an overview of the physico-chemical properties of thermal springs and gaseous emission zones in Nepal, with an emphasis on dissolved inorganic carbon and gaseous CO_2 emissions. We then discuss the spatial organization and earthquake-induced temporal variations of the deep metamorphic carbon release, and identify research perspectives for the coming years.

9.2. Overview of thermal springs geochemistry in Nepal

A total of 227 thermal springs have been studied so far in Nepal (see Volume 1 – Chapter 8 and the associated sites name and number #). Among them, about 20 have enough flow rate (above 0.5 $L.s^{-1}$) and temperature (above 30°C) to be used efficiently by the population, and remain therefore important pilgrimage sites.

Considering the whole data-set, the water temperature has an overall average of 36±1°C and varies from about 7°C to over 77°C (#3 Kermi, far-western Nepal) (Figure 9.1a). Forty-one springs have a temperature above 50°C.

The pH of the hot water (Figure 9.1b) is generally near neutrality, with an overall average of 6.8±0.1, but sometimes shows acidic values below 6 for 17 springs and more rarely basic values above 8 for 12 springs.

The electrical conductivity (Figure 9.1c), with an overall average of 1.7±0.2 $mS.cm^{-1}$, ranges from about 0.01 $mS.cm^{-1}$ to more than 15 $mS.cm^{-1}$ (#35 Bahundanda, western Nepal), depending on the site. Nine springs have an electrical conductivity greater than 5 $mS.cm^{-1}$.

Dissolved inorganic carbon (DIC) content (Figure 9.1d) is generally high, with an overall average of 20±1 $mmol.L^{-1}$, and ranges from 1.3 to to 170 $mmol.L^{-1}$ (Narsing, western Nepal). Seventeen springs have a DIC content greater than 40 $mmol.L^{-1}$. For the carbon isotopic composition of DIC, $\delta^{13}C_{DIC}$, a large range of values is also observed (–20/+14‰).

Large min/max ranges of values are also found for dissolved radon (0.1–900 $Bq.L^{-1}$) and radium concentrations (<3–1,300 $mBq.L^{-1}$).

This large heterogeneity of thermal water characteristics is also observed with major and trace elements. In this section, we present an overview of the results of major ion chemistry and stable isotopic composition analyses of thermal spring waters in the Nepal Himalaya, with an emphasis on dissolved inorganic carbon content.

Despite a large heterogeneity, sometimes shown to be related to lithology with reported differences between the Tethyan Sedimentary Sequence (TSS), the Greater Himalayan Sequence (GHS), the Lesser Himalayan Sequence (LHS) and the Sub-Himalayan Sequence (SHS) in given valleys (e.g. Evans et al. 2004; Ghezzi et al. 2019, and Volume 2), some common characteristics and features emerge from the large dataset, in particular when considering the different regions of Nepal.

The ternary diagram of major cations Mg–Ca–(Na+K) (Figure 9.2a) shows that, in general, the thermal springs are richer in Ca and (Na+K) than in Mg. Moreover, the variation of Ca–Mg content is smaller than the variation of Na+K content. A linear trend is observed between two endmembers: one at $(Na{+}K)_{100\%}$ and the other between $Ca_{50\%}Mg_{50\%}$ and $Ca_{75\%}Mg_{25\%}$. This trend is clear for the western Nepal springs, and especially for the Bahundanda springs. However, it is also observed for the central Nepal springs, particularly for the Syabru-Bensi springs, but they show a larger Mg content on average. Springs of mid-western and eastern Nepal show on average an even larger Mg content. Springs of the MFT zone show large Na+K content.

The ternary diagram of major anions SO_4–Cl–HCO_3 (Figure 9.2b) shows more scattered data. As previously reported (e.g. Evans et al. 2004), the HCO_3 content is large and represents for all the studied thermal springs more than 70% of the total dissolved anionic charge. Generally, low Cl content is found for springs of central Nepal, while extremely large Cl content is found for numerous springs of western Nepal, especially for the Bahundanda springs. All springs of eastern Nepal and the two springs from far-western Nepal are concentrated around the pole $10SO_{4(5\%)}-10Cl_{70\%}-HCO_{3(20\%)}$. Springs of the MFT zone are concentrated around the pole $10SO_{4(20\%)}-10Cl_{15\%}-HCO_{3(65\%)}$. Springs of mid-western Nepal show larger Cl content on average. These composition differences suggest along-strike heterogeneity of the main lithology at a large spatial scale in the MCT zone throughout Nepal. Conversely, as integrator of the crossed rock, thermal springs may be precious indicators of the lithology variations along-strike, accessing parameters almost impossible to infer otherwise such as the volume of a given lithology crossed.

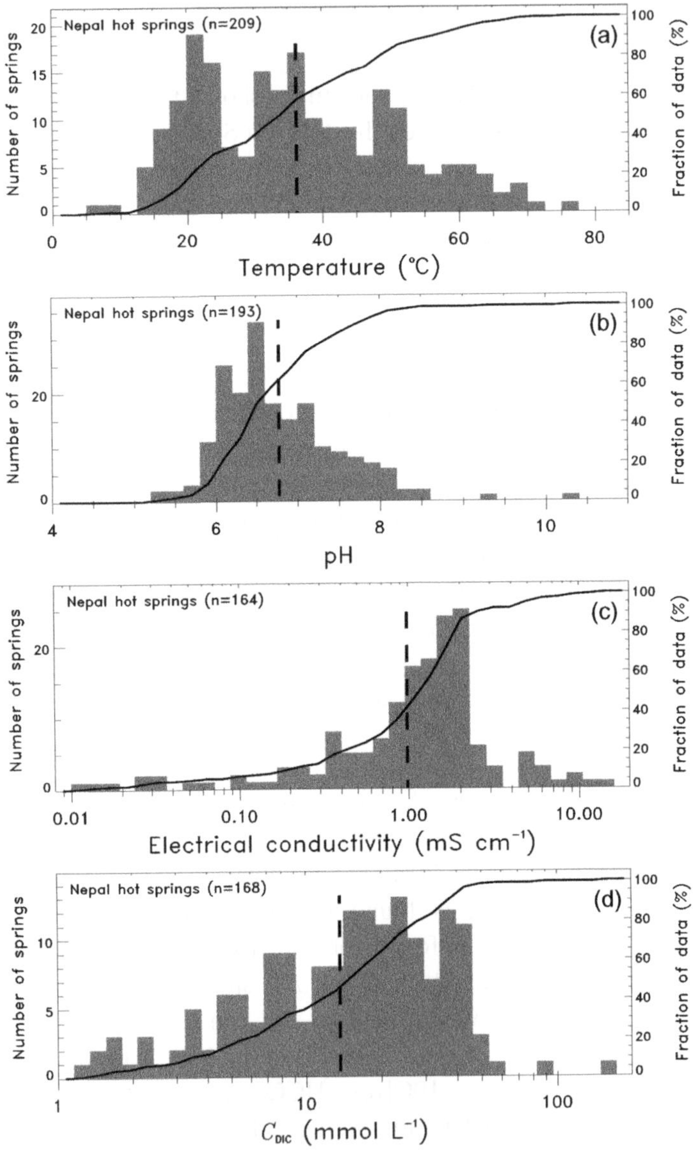

Figure 9.1. *Distributions of water (a) temperature, (b) pH, (c) electrical conductivity and (d) dissolved inorganic carbon (DIC) content of thermal springs in the Nepal Himalaya*

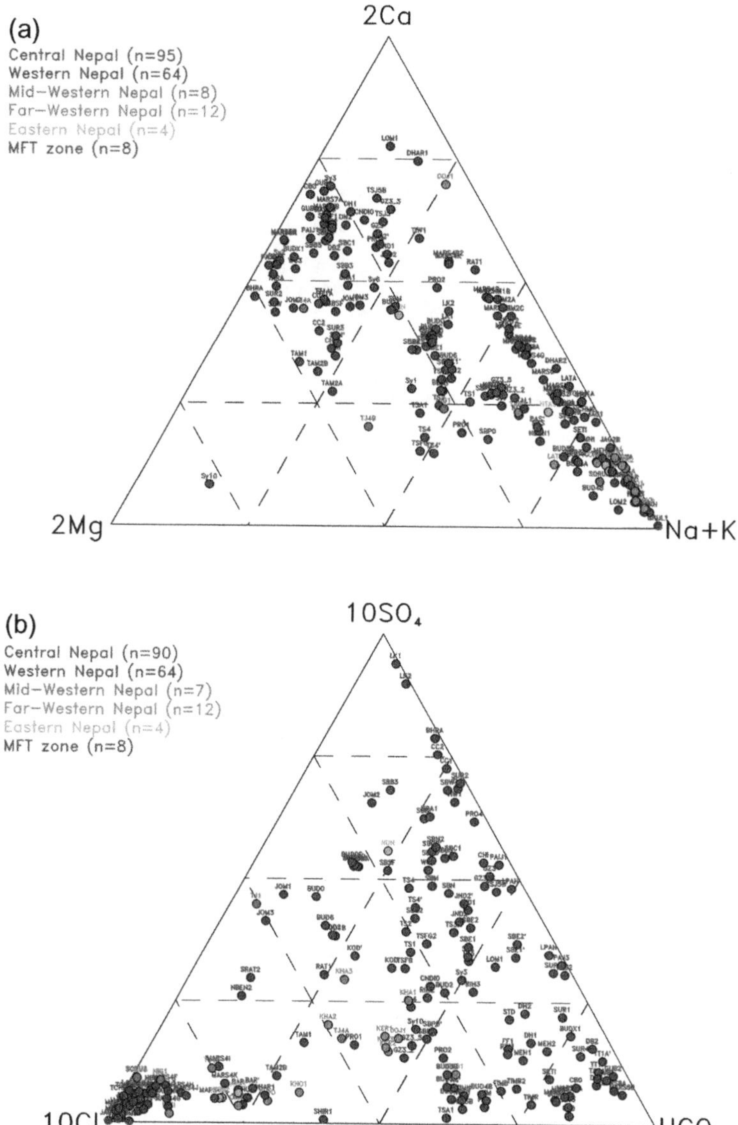

Figure 9.2. *(a) Mg–Ca–(Na+K) and (b) SO_4–Cl–HCO_3 ternary diagrams for thermal spring waters in the Nepal Himalaya. For a color version of this figure, see www.iste.co.uk/cattin/himalaya3.zip*

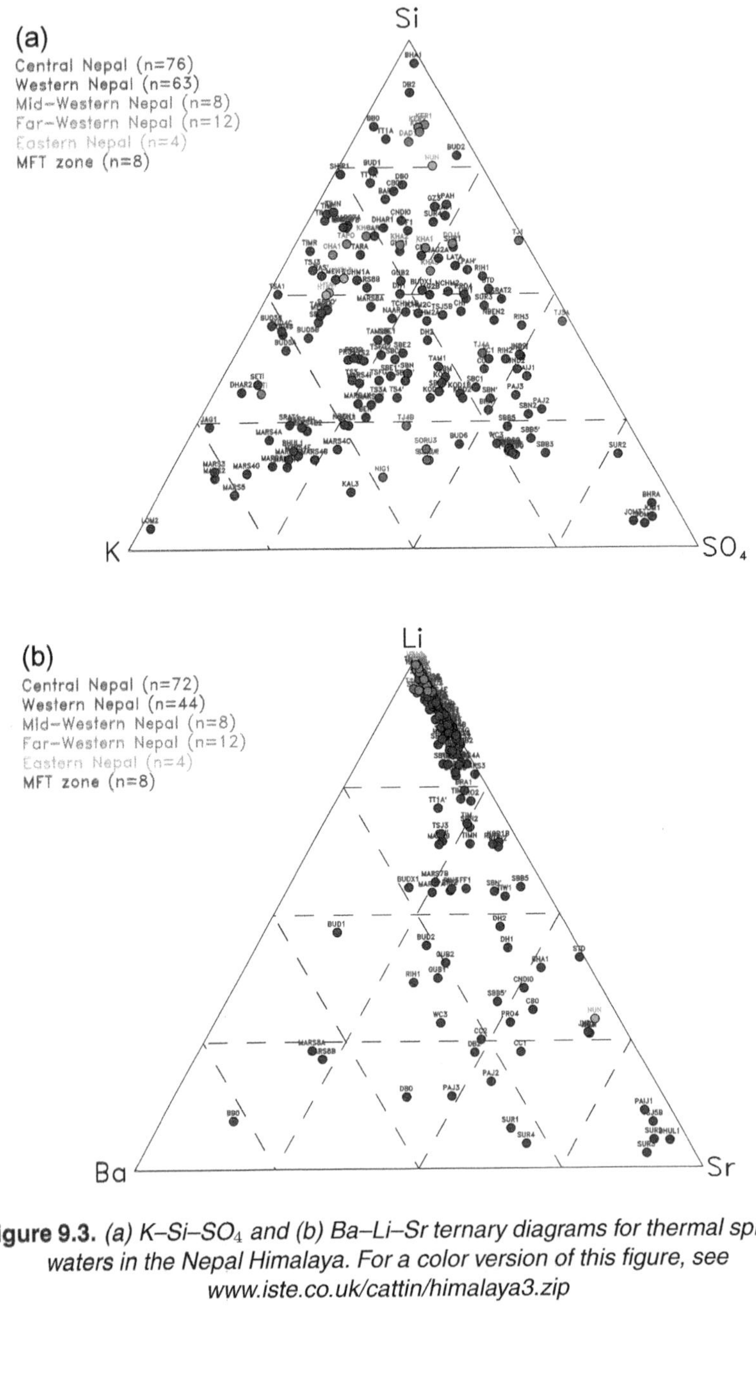

Figure 9.3. *(a) K–Si–SO_4 and (b) Ba–Li–Sr ternary diagrams for thermal spring waters in the Nepal Himalaya. For a color version of this figure, see www.iste.co.uk/cattin/himalaya3.zip*

The ternary diagram of major ions K–Si–SO_4 (Figure 9.3a) also shows scattered data. In general, SO_4 content remains low in most springs of western, eastern and far-western Nepal and higher in most springs of central Nepal; intermediate SO_4 content is found for springs of mid-western Nepal and the MFT zone. Larger K content is observed for springs of western Nepal compared with those of central Nepal and the MFT zone; intermediate K content is found for springs of eastern and mid-western Nepal. A large Si content is obtained for several springs of central and western Nepal, but the values are largely scattered. Most springs of the Na–K–Cl type are found in western Nepal, while most springs of the Ca–Mg–SO_4 type are found in central Nepal. In mid-western, far-western and eastern Nepal, springs are mostly of the Na–K–Cl type, while it is less clear due to the relatively small number of available data. In the MFT zone, springs are generally of the Na–HCO_3 type.

The ternary diagram of Li–Ba–Sr (Figure 9.3b) shows that most springs have a large Li content. Most of the springs with a lower temperature have a smaller Li content. The springs of the MFT zone show a lower Li content and a higher Sr content. The large alkaline Na–K–Li contents at high temperature springs suggest strong alteration at high temperature.

All of these springs contribute to the alkalinity charge of the main Himalayan rivers flowing through Nepal ((Evans et al. 2004), but they are not enough to explain the large content of some cations (Na, K) found in these rivers. Several unidentified seepages of thermal springs are thus strongly suspected on the riverbeds at all places of the MCT zone. Furthermore, the large Na content of thermal springs of the MFT zone cannot entirely explain the large Na content of Siwalik rivers (Surai and Rapti rivers) (Evans et al. 2004), also suggesting numerous hidden thermal water discharges below the river level.

The δD and $\delta^{18}O$ values of thermal spring waters (Figure 9.4a) show a clear linear relationship that follows the local meteoric water line (Garzione et al. 2000; Gajurel et al. 2006), as evidenced in the Nepal Himalaya in previous studies (e.g. Grabczak and Kotarba 1985; Becker et al. 2008; Evans et al. 2008). Here, based on the available dataset, slight differences can be observed between western and central Nepal, suggesting that a specific meteoric water line should be defined in each valley. Nevertheless, all the studied thermal springs follow the meteoric water line and are thus of meteoric origin. Almost all δD values of thermal springs (Figure 9.4b) fall a little below the master curve for meteoric water obtained in the Kali Gandaki valley (Garzione et al. 2000), which suggests that, generally, the altitude of meteoric water recharge is slightly higher than the altitude of the thermal spring. However, some springs with the same altitude can have a different altitude of water recharge, as observed, for example, in central Nepal.

The available data on thermal springs in Nepal of helium isotopic ratio normalized by that of the atmosphere, R/R_A, albeit scarce so far (n=6), show small values, mostly below 0.2. As shown in previous studies (Marty et al. 1996; Becker 2005; Girault 2011), with negligible mantellic ^{3}He and large radiogenic ^{4}He contents, these results suggest a pure crustal radiogenic source for dissolved helium in thermal springs in the Nepal Himalaya.

When thermal spring data of δ^{13}C of DIC is plotted as a function of DIC concentration (C_{DIC}) (Figure 9.5a), a general trend is observed showing an increase of $\delta^{13}C_{DIC}$ with C_{DIC}. Most of the thermal waters where bubbling or nearby CO_2 degassing zones are observed show $\delta^{13}C_{DIC}$ above 0‰, suggesting degassing from the hydrothermal water. In mid-western Nepal, C_{DIC} is smaller. In eastern Nepal, $\delta^{13}C_{DIC}$ values are around 0‰. The other thermal springs show smaller $\delta^{13}C_{DIC}$ and C_{DIC} values. In the MCT zone, thermal springs have a smaller $\delta^{13}C_{DIC}$ (≈ -10‰) and intermediate C_{DIC} ($\approx$ 10 mmol.L^{-1}).

Following previous works in Italy (Frondini et al. 2019; Chiodini et al. 2020), we calculated for each spring (Figure 9.5b) the carbon content from the sources external from thermal waters (C_{ext}) by subtracting the carbon coming from the dissolution of carbonate minerals to C_{DIC} values. Similarly, $\delta^{13}C_{ext}$ is calculated based on the mean δ^{13}C value for carbonates (–2.0±0.1 ‰; Evans et al. 2008), and 1/C_{ext} versus $\delta^{13}C_{ext}$ follow a linear relationship, which confirms that, in all the thermal springs in the Nepal Himalaya, part of the carbon is removed during their ascent through degassing and calcite precipitation. These processes imply carbon isotopic fractionation (Becker et al. 2008; Evans et al. 2008; Girault et al. 2014a). These observations also suggest that, due to the large amount of degassed CO_2 from the roots of hydrothermal systems to the shallow surface, before the exit of the thermal springs, an important part of the degassed CO_2 will potentially reach the surface, mainly leading to surface manifestations such as bubbling or gaseous degassing zones.

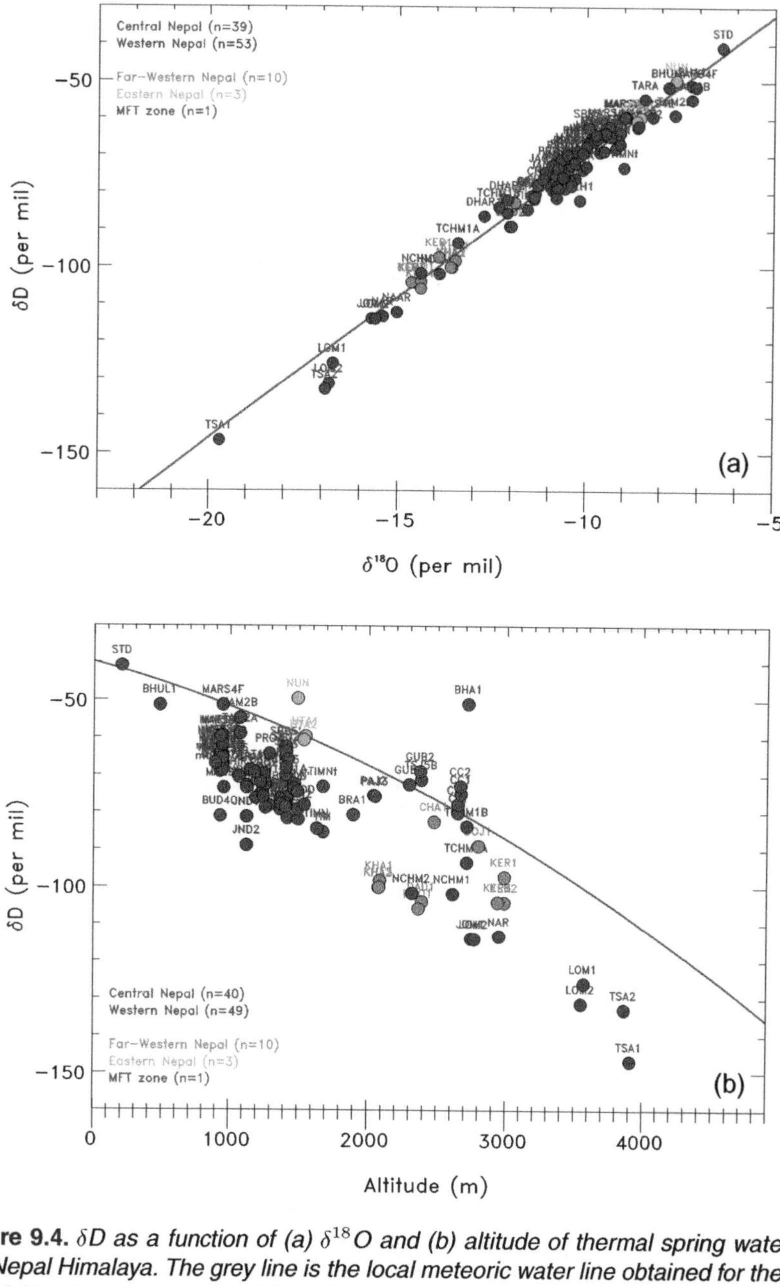

Figure 9.4. *δD as a function of (a) $\delta^{18}O$ and (b) altitude of thermal spring waters in the Nepal Himalaya. The grey line is the local meteoric water line obtained for the Kali Gandaki valley, western Nepal (Garzione et al. 2000), which depends on the altitude. For a color version of this figure, see www.iste.co.uk/cattin/himalaya3.zip*

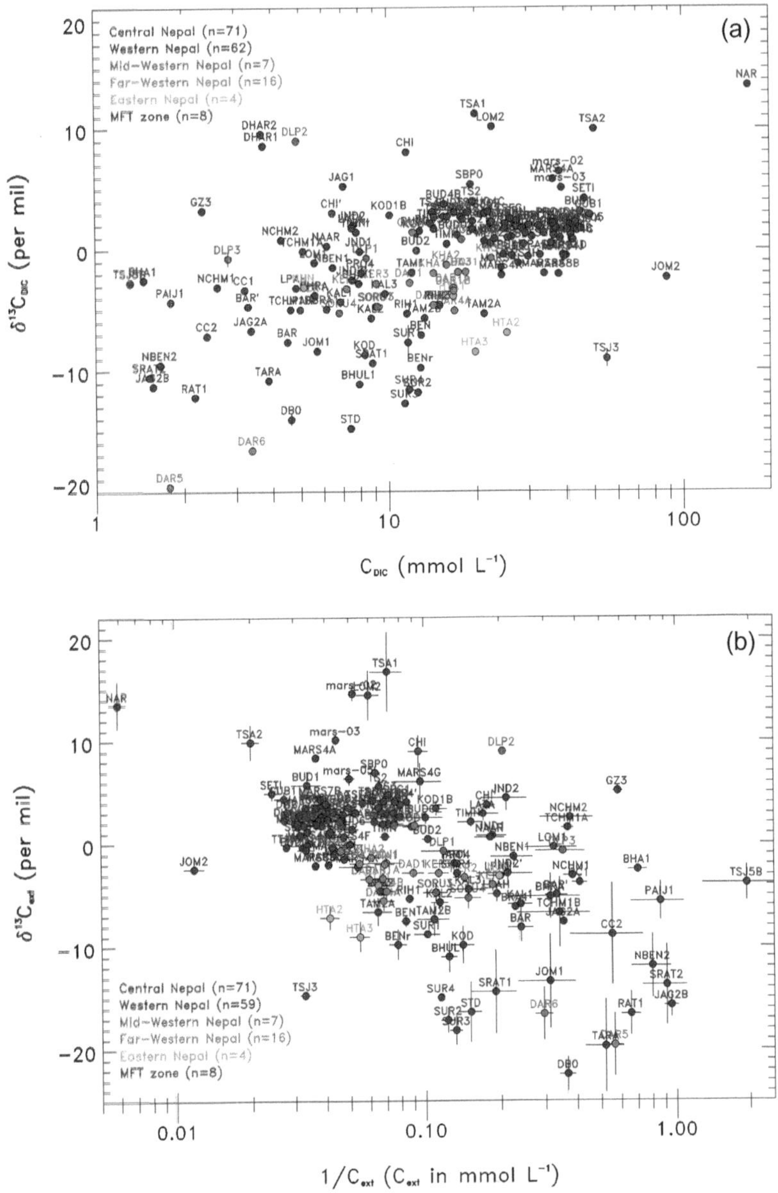

Figure 9.5. *Dissolved inorganic carbon characteristics of thermal spring waters in the Nepal Himalaya: (a) C_{DIC} versus $\delta^{13}C_{DIC}$ and (b) calculated $1/C_{ext}$ versus $\delta^{13}C_{ext}$. Calculations of C_{ext} and $\delta^{13}C_{ext}$ in (b) are described in the text and based on the literature (Frondini et al. 2019; Chiodini et al. 2020). For a color version of this figure, see www.iste.co.uk/cattin/himalaya3.zip*

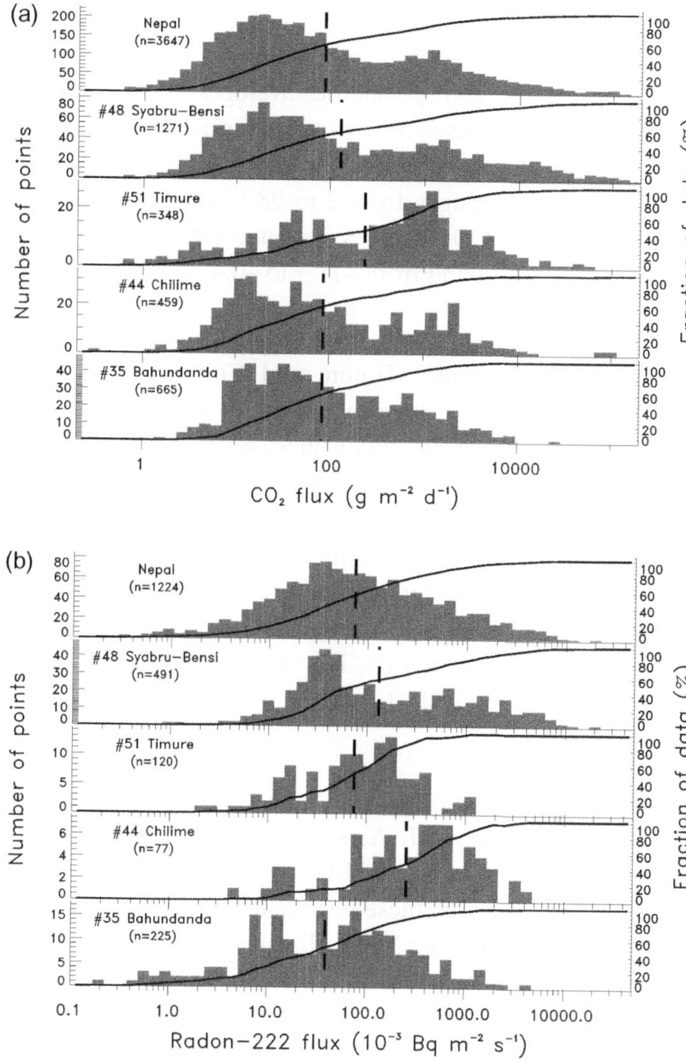

Figure 9.6. *Distributions of (a) CO_2 flux and (b) radon-222 flux in the Nepal Himalaya*

9.3. Overview of gaseous emission zones in Nepal

A total of 40 sites have been studied so far in Nepal in terms of gaseous emissions (see Volume 1 – Chapter 8). Eleven sites show particularly strong degassing in the form of "tectonic" fumaroles and diffuse degassing structures. Considering the whole dataset, with overall arithmetic and geometric averages of 2,400±200 and 87.1±0.3 g.m^{-2}.d^{-1}, respectively, mean surface CO_2 fluxes (n=3,647) range from 0.3 to 340,000 g.m^{-2}.d^{-1} (Figure 9.6a). With overall arithmetic and geometric averages of 420±30×10^{-3} and 74±1×10^{-3} Bq.m^{-2}.s^{-1}, respectively, mean surface radon fluxes (n=1,224) range from 0.2×10^{-3} to $20,000 \times 10^{-3}$ Bq.m^{-2}.s^{-1}. The CO_2 concentration of the gaseous emissions ranges from 2 to 98 % (n=105). The mean carbon isotopic compositions of CO_2 bubbles in water and gaseous CO_2 emissions range from –11.2 to –1.9‰ and from –11.7 to –0.4‰, respectively (n=128).

The distributions of CO_2 fluxes in Nepal and at the four main studied sites in central (Syabru-Bensi, Timure and Chilime) and western Nepal (Bahundanda) (Figure 10.6a) range over more than five orders of magnitude and show two modes: one at low flux values ($\approx$ 20 g.m^{-2}.d^{-1}) and the other at larger flux values ($\approx 1,000$ g.m^{-2}.d^{-1}). At Syabru-Bensi, 99 CO_2 flux values (approximately 8%) are above 10,000 g.m^{-2}.d^{-1}. Although the number of data point is smaller than for CO_2 fluxes, the distributions of radon fluxes in Nepal and at the four main studied sites in central (Syabru-Bensi, Timure and Chilime) and western Nepal (Bahundanda) (Figure 9.6b) show two modes: one main mode at low flux values ($\approx 35 \times 10^{-3}$ Bq.m^{-2}.s^{-1}) and the other at larger flux values (between $\approx 100 \times 10^{-3}$ and $\approx 1,000 \times 10^{-3}$ Bq.m^{-2}.s^{-1} depending on the site).

The available R/R_A data on gaseous emissions in Nepal (n=8) show small values, with mean of 0.06±0.01. As shown in previous studies (Marty et al. 1996; Becker 2005; Girault et al. 2014a), with negligible mantellic ^{3}He and large radiogenic ^{4}He contents, these results indicate a pure crustal radiogenic source for gaseous helium in the Nepal Himalaya.

The $\delta^{13}C_{gas}$ is plotted as a function of $\delta^{13}C_{DIC}$ for all the studied sites (Figure 9.7a). Most of the $\delta^{13}C_{gas}$ values from central and western Nepal, scattered between the organic source (≈ -25‰) and the metamorphic source (≈ 0‰), are closer to the metamorphic source. Only a few data show a mixing of the two sources with mean value of ≈ -11‰ (Lower Dolpo and Tama Koshi areas). When $\delta^{13}C_{gas}$ indicates a metamorphic source, values of $\delta^{13}C_{DIC}$ are above 0‰, suggesting strong CO_2 degassing from hydrothermal water. At each site, $\delta^{13}C_{bubbles}$ show smaller values than $\delta^{13}C_{gas}$. This can be explained by a two-step fractionation process (Becker 2005; Girault et al. 2014a). Large CO_2 fluxes are associated with $\delta^{13}C_{gas}$ over –4‰ (Figure 9.7b), while smaller CO_2 fluxes, compatible with background values, show $\delta^{13}C_{gas}$ of ≈ -11‰.

Importantly, the association of large CO_2 fluxes with high $\delta^{13}C_{gas}$ values suggests that on the ground surface, we may have access to a gas that was only slightly fractionated during its ascent from its source at crustal depth.

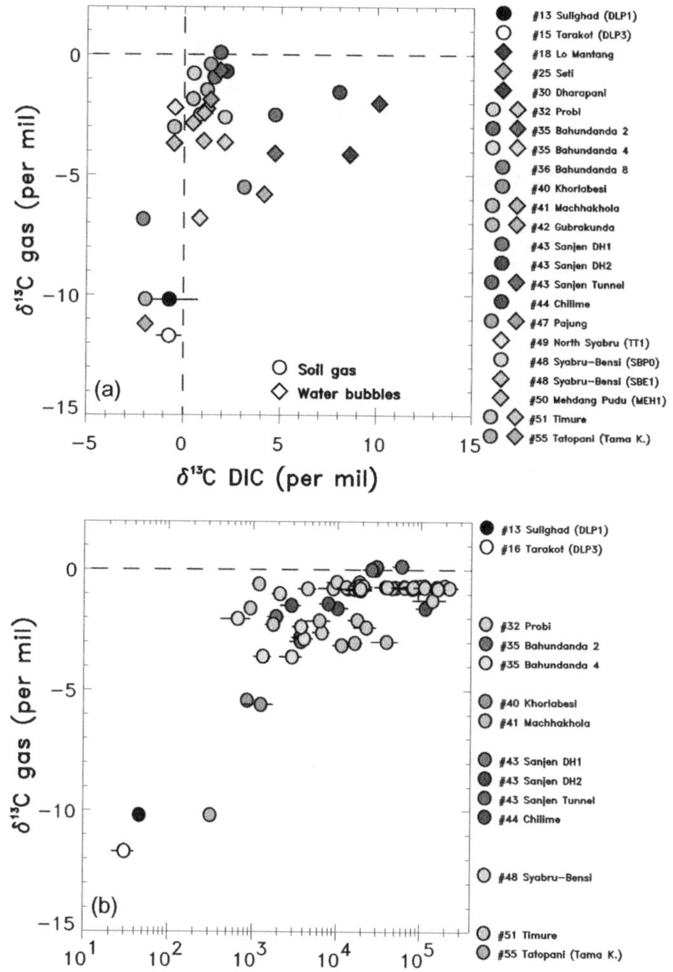

Figure 9.7. *Carbon isotopic composition of gaseous CO_2 emissions in the Nepal Himalaya: (a) $\delta^{13}C_{gas}$ as a function of $\delta^{13}C_{DIC}$ and (b) $\delta^{13}C_{gas}$ as a function of gaseous CO_2 flux. In (a), data from soil gas and water bubbles are shown as circle and diamond, respectively. For a color version of this figure, see www.iste.co.uk/cattin/himalaya3.zip*

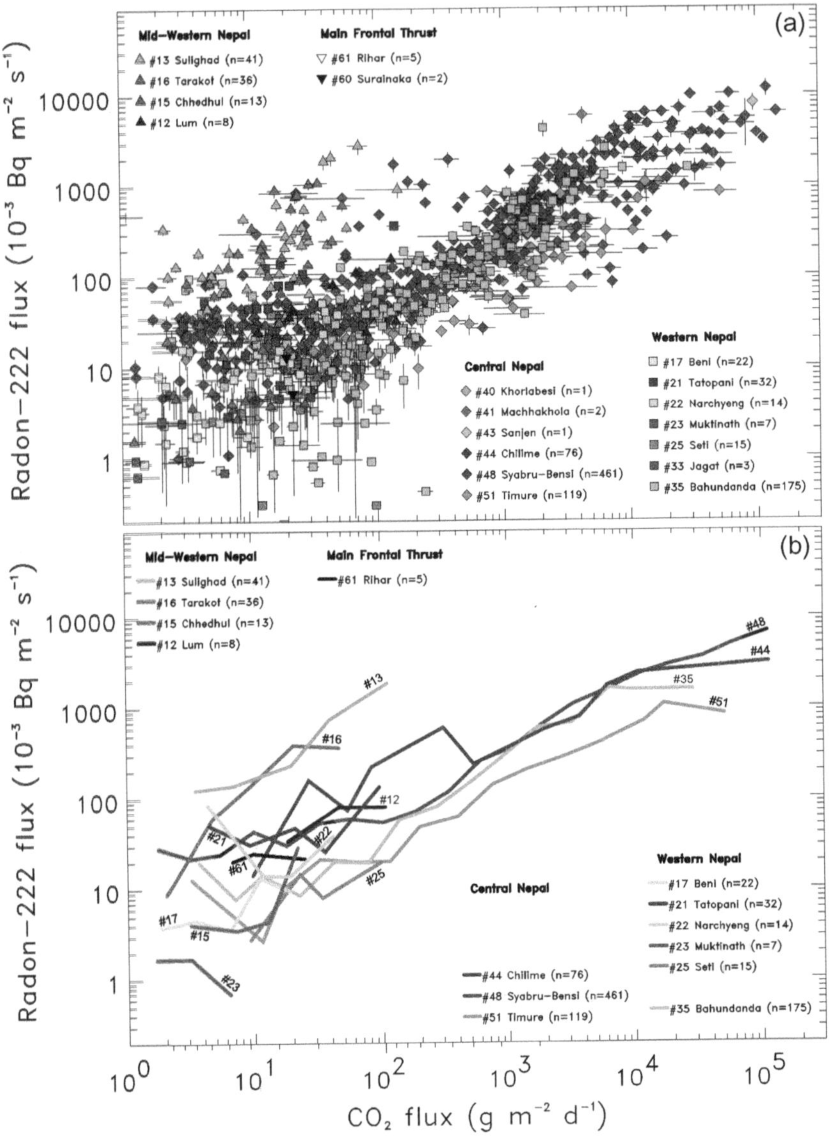

Figure 9.8. *Overview of gaseous CO_2 flux versus radon-222 flux in the Nepal Himalaya: (a) surface fluxes data and (b) average of the data. For a color version of this figure, see www.iste.co.uk/cattin/himalaya3.zip*

The CO_2 flux – radon flux correlation for all sites (Figures 9.8a,b) shows a striking relationship over more than five orders of magnitude. In central and western Nepal, the fluxes reach exceptionally large values, suggesting radon is carried by CO_2 to the surface, as shown in previous studies (Girault et al. 2009, 2014a,b). Based on coupled CO_2 and radon transport modeling, this correlation suggests an advective transport of CO_2 carrying radon with velocities up to the order of 0.1 to 1 m.s^{-1} (Girault and Perrier 2014), and reliably constrains the depth of CO_2 degassing (Girault et al. 2022). In mid-western Nepal, larger radon fluxes are observed for smaller CO_2 fluxes, suggesting a stronger radon source term near the surface (Girault et al. 2014b). In the MFT zone, fluxes are compatible with background values.

To summarize, based on this review of thermal springs and gaseous emissions and on previous studies (e.g. Becker et al. 2008; Evans et al. 2008; Derry et al. 2009; Girault et al. 2014b, 2018c, 2022; Groppo et al. 2022), we can propose the following conceptual model for CO_2 in the Nepal Himalaya. The gaseous CO_2 is produced at more than 5 km depth by metamorphic reactions including decarbonation, and is transported from its source by fluid immiscibility during prograde metamorphism. Then, it percolates through the dense fracture and fault networks of the MCT, MBT and MFT zones and forms gas reservoirs at crustal depths. Infiltrated meteoric water mixes with metamorphic CO_2 at the roots of hydrothermal systems where the resulting hydrothermal fluid is heated up by the high local thermal gradient. CO_2 is then degassed from the hydrothermal water at a depth of 100 m or more at temperatures between 70°C and 120°C. Finally, CO_2 can cross near surface aquifers with little or no fractionation before reaching the surface where various hydrothermal manifestations such as water bubbling, "tectonic" fumaroles and diffuse degassing structures occur.

9.4. Spatial organization of crustal fluid release

From the extensive dataset presented above, we get a robust overview of the along-strike variations of crustal fluid emissions.

In a given region of the MCT zone, such as the upper Trisuli valley, sites separated by 5–10 km show similar characteristics of the crustal fluid release. This suggests that hydrology and topography locally are not the predominant controlling factors of the small-scale spatial heterogeneity of the CO_2 release. Instead, in such a given valley, the consistent values of $\delta^{13}C_{gas}$ and maximum CO_2 flux indicate a similar metamorphic CO_2 source at crustal depths (Girault et al. 2014b). However, when considering the whole available dataset from far-western to eastern Nepal, significant differences are observed along the Nepal Himalayan belt.

Several aforementioned parameters of the thermal springs and gaseous emissions are plotted after projection against the Nepal Himalayan arc (Figure 9.9). Larger $\delta^{13}C_{gas}$, including gaseous emission zones and water bubbling, are found in central and western Nepal from about 400–580 km distance along the arc. The values of $\delta^{13}C_{DIC}$ show a larger variation in western and central Nepal, with, albeit less visible, larger values in the central part of the projection. The largest DIC concentrations are also found between about 400 and 580 km distance along the arc. Thermal spring temperature shows large differences along-strike, with larger temperature in central, western and far-western Nepal, and a progressive decrease towards the east. The maximum pH values are found in the central part of the projection. The largest electrical conductivity values are observed in western and central Nepal. The dissolved radium concentration appears relatively large in several valleys of Nepal compared to other sites in similar context (Girault et al. 2018a). The dissolved radon concentration is significantly higher on average in mid-western and central Nepal.

The dissolved radon discharge, estimated by the product of radon concentration and water flow rate, shows a progressive decrease from west to east, with the exception of the Trisuli valley in central Nepal. The gaseous radon discharge, estimated by kriging and interpolation of the radon flux dataset at each site, shows large values in central Nepal. In total, summing all the radon output, we obtain a total radon discharge in the MCT zone of 7.8 ± 1.6 kBq.s^{-1}, including about 43% from the gaseous emission zones.

The DIC discharge, estimated by the product of DIC concentration and water flow rate, appears slightly higher in the central part of the Nepal Himalayan arc. The gaseous CO_2 discharge, estimated by kriging and interpolation of the CO_2 flux dataset at each site, shows large values in central and western Nepal, with about three orders of magnitude larger values than those obtained in mid-western and eastern Nepal. This central zone with larger total gaseous CO_2 discharge, between about 420 and 580 km distance along the arc, also depicts the largest $\delta^{13}C_{gas}$. In total, summing all the estimated carbon output, we obtain a total CO_2 discharge in the MCT zone of 11.5 ± 2.8 mol.s^{-1}, including about 96% from the gaseous emission zones.

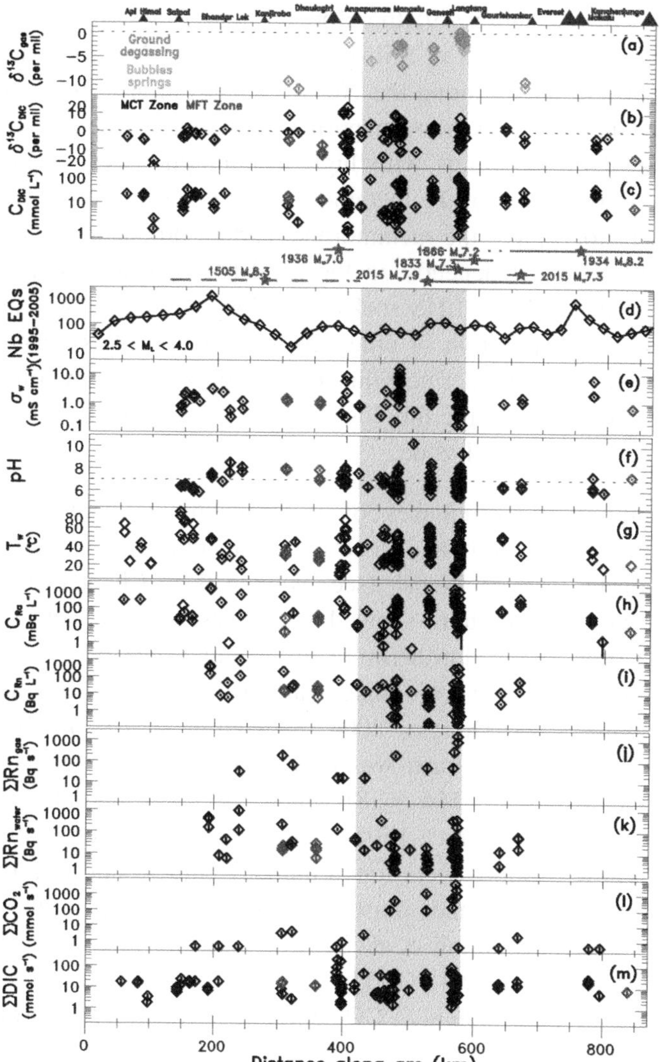

Figure 9.9. *Spatial variations of geothermal system characteristics along the Nepal Himalayan arc. Data from the MCT and MFT zones are plotted in black and blue, respectively. The number of earthquakes comes from the 1994 to 2014 NSC–DASE catalogue. Epicenters and rupture lengths of the 1505 and 1934 mega-earthquakes, the 1833, 1866 and 1936 intermediate earthquakes, and the 2015 Gorkha earthquake are shown (see Volume 3 – Chapters 5 and 8). For a color version of this figure, see www.iste.co.uk/cattin/himalaya3.zip*

Our overview (Figure 9.9) therefore clearly shows, pending more data on the degassing observed at Sribagar (#1, far-west Nepal), that a segment of about 160 km length along the Nepal Himalayan arc, from about 420 to 580 km distance between the Annapurnas and Langtang ranges, concentrates the release of crustal fluids with high metamorphic CO_2 degassing. This large-scale organization of the metamorphic CO_2 release might be controlled by several factors. First, the metamorphic carbon production rate may affect the CO_2 discharge. It depends on the available rock material, such as the thickness of the GHS or LHS rocks, the along-strike changes in lithology, the temperature distribution or other rheological factors associated with the MCT zone. Second, the permeability of the upper crust may affect the CO_2 discharge. It depends on the lithology and on the occurrence and state of the fault networks of the MCT zone.

The number of earthquakes of local magnitude between 2.5 and 4 appears small in western and central Nepal compared with mid-western, eastern and far-western Nepal (see Volume 1 – Chapter 4). The high CO_2 emitting segment thus showed the lowest seismic activity over the 1994–2014 time period. Furthermore, it is sandwiched between the surface rupture of the 1505 and 1934 mega-earthquakes (Ambraseys and Jackson 2003; Ambraseys and Douglas 2004; Sapkota et al. 2013). The high CO_2 emitting segment, by contrast, was located in the seismic gap with smaller major earthquakes ($7 \leq Mw \leq 8$) which have only partially released the accumulated slip deficit since 1505 (Gutenberg and Richter 1949; Szeliga et al. 2010). These relations were found unlikely coincidental in 2014, because Himalayan mega-earthquakes must have a dramatic effect on the large-scale crustal permeability (Girault et al. 2014b). This inference was unfortunately confirmed by the deadly 2015 Gorkha earthquake that hit central Nepal, with a rupture covering the whole high CO_2 emitting segment (Adhikari et al. 2015; Galetzka et al. 2015).

9.5. Temporal variations of crustal fluid release: a tectonic control

The aforementioned observations lead to the concept of dynamic permeability of the upper crust (Manning and Ingebritsen 1999; Ingebritsen and Manning 2010; Manga et al. 2012; Miller 2020) related to the Himalayan seismic cycle, affecting the release of crustal fluid. In addition, we have sufficient data over long periods of time to discuss key examples of the temporal variations of crustal fluid emissions triggered by the 2015 Gorkha earthquake. Here we focus on two hydrothermal systems of central Nepal (Syabru-Bensi and Chilime) that were severely influenced by the 2015 Gorkha earthquake.

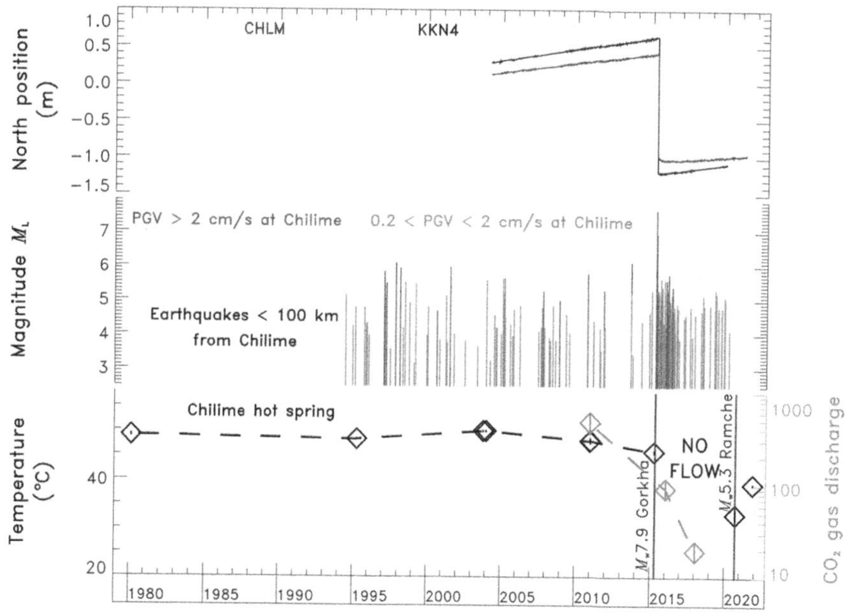

Figure 9.10. *Temporal variations of the geothermal system of Chilime, central Nepal. From top to bottom are shown as a function of time: corrected northward position of the GPS stations CHLM and KKN4 (UNAVCO database), local magnitude (green and purple) of earthquakes (Adhikari 2021; Adhikari et al. 2022) $\leq$ 100 km from Chilime with vertical Peak Ground Velocity (PGV) above 0.2 cm.s^{-1} at Chilime (NSC–DASE earthquake catalogue), and temperature of the Chilime hot spring water and gaseous CO_2 discharge of the Chilime site. The hot spring stopped flowing after the 2015 Mw 7.9 Gorkha earthquake and started to flow again intermittently following the September 15, 2020 Mw 5.3 Ramche earthquake. For a color version of this figure, see www.iste.co.uk/cattin/himalaya3.zip*

The available time series of water temperature and gaseous CO_2 discharge obtained at Chilime, central Nepal (Figure 9.10) reveal an impressive connection to the seismic cycle. The record of Chilime hot spring temperature is the longest with the first reliable measurement done in 1980 (Grabczak and Kotarba 1985). Then, other measurements of temperature were performed in 1995 (Evans et al. 2004), 2000 (Perrier et al. 2002), 2003 (Becker 2005), 2004 (Perrier et al. 2009), 2011 (Girault et al. 2014b, 2018b), 2015 (Girault et al. 2018c), 2020 and 2021. Water temperature remained stable around

48.9±0.4°C and with a large flow rate of 5.0±0.2 L.s^{-1} before the Gorkha earthquake. About 20 m above the spring mouth, the gaseous emission zone released CO_2 in 2011 with a discharge of 470±120 mmol.s^{-1} (Girault et al. 2018b). After the 2015 Gorkha earthquake, that shook the area with an estimated vertical peak ground velocity (PGV) over 7 cm.s^{-1} (Adhikari 2021), the hot spring flow became intermittent and suddenly stopped about six months after the main shock (Girault et al. 2018c). Significant changes were also recorded at the gaseous emission zone which showed a progressive decrease in CO_2 discharge with 91 ± 19 mmol.s^{-1} in 2016 and 19 ± 4 mmol.s^{-1} in 2018 (Girault et al. 2018c). The absence of thermal water flow lasted about five years, until warm water flew again in 2020 at the same place, with albeit a smaller temperature and a smaller flow rate. The reactivation of the water flow coincided with the Mw 5.3 Ramche earthquake which, in 2020, strongly hit the area. These observations suggest that the hydrothermal unrest at Chilime, including the thermal spring cessation and the decrease in gaseous CO_2 discharge, was triggered by the Gorkha earthquake crisis (Girault et al. 2018c). The Chilime hydrothermal system is thus in near-critical condition, and the effects of the Gorkha earthquake crisis are not over, sustaining hydrothermal unrest at this site more than six years after the seismic event.

The available time series of water temperature and DIC concentration of the main hot spring, and surface CO_2 flux obtained at Syabru-Bensi, central Nepal (Figure 9.11) confirm the previous conclusion. The water temperature time series from 2003 to 2019, the DIC content and the CO_2 flux time series over the period from 2007 to 2020 are shown (Perrier et al. 2002, 2009; Evans et al. 2004, 2008; Becker 2005; Byrdina et al. 2009; Girault and Perrier 2014; Girault et al. 2014b, 2018c). Water temperature, stable at 60.7 ± 0.1°C for more than 12 years before the Gorkha earthquake, increased to 64.1 ± 0.3°C after the earthquake in winter 2015. After a peak about one year after the mains hock, water warming persisted more than two years after the mainshock, and then decreased to pre-earthquake values in 2018. Temperature remained similar to pre-earthquake values, but was unstable in 2018 and 2019. The DIC content also showed an increase following the earthquake, then followed by large variations in 2018 and after. The CO_2 flux, measured at two locations in 2007, showed a clear increase following the earthquake and a peak in 2016. The values remained two orders of magnitude higher after. Based on earthquake and GPS records, these observations suggest that the hydrothermal unrest at Syabru-Bensi, including the increases of water temperature, DIC content and CO_2 flux, was also triggered by the Gorkha earthquake crisis

(Girault et al. 2018c). The Syabru-Bensi hydrothermal system is thus also in near-critical condition, unstable and potentially able to diverge unpredictably.

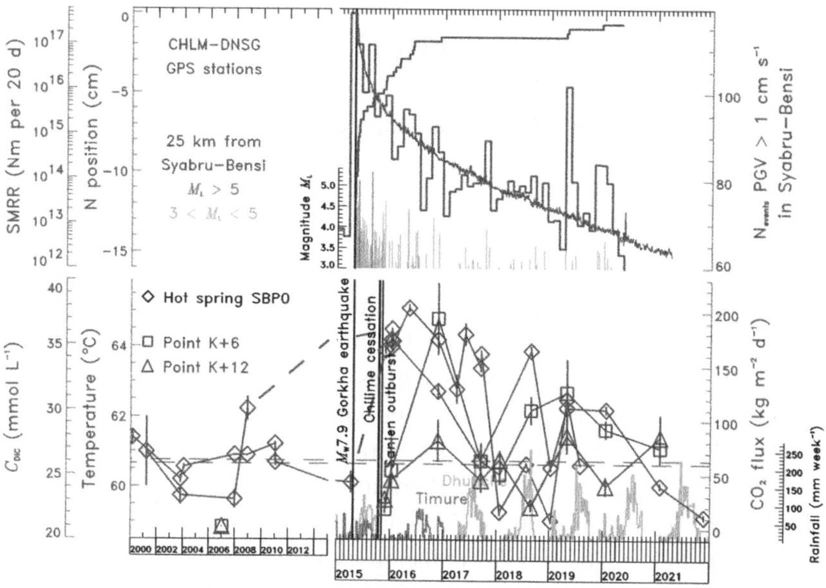

Figure 9.11. *Temporal variations of the geothermal system of Syabru-Bensi, central Nepal. Top plot: corrected northward position difference of GPS stations CHLM minus DNSG (red) (UNAVCO database), cumulated number of aftershocks with vertical Peak Ground Velocity (PGV) above 1 cm.s^{-1} in Syabru-Bensi (blue), Seismic Moment Release Rate (SMRR) (orange) and local magnitude (green and purple) of earthquakes (Adhikari 2021; Adhikari et al. 2022) $\leq$ 25 km from Syabru-Bensi (NSC–DASE earthquake catalogue). Bottom plot: water temperature (red) and DIC concentration (orange) of the main hot spring of Syabru-Bensi (SBP0), surface CO_2 fluxes (blue) at two locations on the alluvial terrace above the hot spring (K+6 and K+12) and the rainfall data in Dhunche, 7 km southwest of Syabru-Bensi (green) and Timure, 9 km to the north (purple) (courtesy of Nepal Department of Meteorology, Kathmandu). For a color version of this figure, see www.iste.co.uk/cattin/himalaya3.zip*

While the Chilime hot spring cessation about six months after the main shock could result from hydrogeological fluid diffusion, the increase in water temperature and gaseous CO_2 discharge reported at Syabru-Bensi could be explained by a doubling of the crustal permeability (Girault et al. 2018c). This increase in vertical permeability is compatible with the observation of post-seismic afterslip (Gualandi et al. 2017). This particularly long duration of post-seismic CO_2 release, rarely observed in a large orogen (Zheng et al. 2013; Geersen et al. 2016), may also be explained by significant changes in normal stress (Bonini 2022). These observations have revealed that the crustal fluids released at hydrothermal systems in the Nepal Himalaya are sensitive to co-, post- and possible pre-seismic deformation (Girault et al. 2018c). They have also suggested that the release of metamorphic CO_2 from active orogen is non-stationary. The monitoring of key parameters at hydrothermal systems in the Nepal Himalaya could thus be of great interest to record the temporal dynamics of crustal permeability at an unprecedented large spatial scale.

9.6. Conclusion

In this chapter, based on the dataset described in Volume 1 – Chapter 8, we have presented an overview of the current knowledge on hydrothermal systems in Nepal. The metamorphic CO_2 emission in the Nepal Himalaya, as dissolved in water or as gaseous discharges, is spatially organized. A massive CO_2 emission is concentrated along a 160-km-long segment of the Himalayan arc in the MCT zone between the Annapurnas and Langtang ranges. Temporal variations of CO_2 emissions and thermal spring characteristics have been found to strongly depend on the Himalayan seismic cycle (see Volume 3 – Chapters 5, 6 and 8). The significant increases and decreases of CO_2 flux on the ground and the variations of thermal spring water chemistry and flow rate, up to a complete cessation in central Nepal following the 2015 Gorkha earthquake, are clear indicators of the high sensitivity to earthquakes of the hydrothermal systems in the Nepal Himalaya. Several years after the earthquake crisis, only a few hydrothermal systems have returned to pre-seismic observations, and most hydrothermal parameters remain highly variable, suggesting still an important earthquake-induced instability. This also suggests that, while some sites were strongly impacted and may not completely recover, other sites have a higher resilience to large earthquakes, and may have experienced several large earthquakes in their entire lifetime. Hydrothermal systems on hillslope or thermal springs near the river bank can also be dramatically modified by large landslides or exceptional floods (see Volume 3 – Chapters 3 and 4).

To advance in our understanding of Nepal hydrothermal systems, first, a crucial field exploration effort is needed in unexplored regions of the MCT zone such as eastern Nepal, far-western Nepal, and in the MFT and MBT zones. Indeed, a large amount of CO_2 discharge may have escaped measurement so far. Other exploration methods such as sensors measuring surface temperature and CO_2 air concentration installed on board aircrafts or drones, or other yet unidentified proxies, are needed to get an estimate of CO_2

emissions at a large spatial scale over this rough terrain. Second, beyond the local economic value of hot springs as touristic and religious locations (Rai et al. 2020), the potential for geothermy, only briefly envisaged to date (Ranjit 2015), may be tested at selected hydrothermal sites based on a dedicated operational drilling program. Furthermore, exploration should be extended to other locations outside Nepal, in particular in the MCT zone such as in India and Bhutan (Singh et al. 2004; Tiwari et al. 2016), to obtain a general overview of the hydrothermal systems and CO_2 emissions at the scale of the whole Himalayan range. Once the current CO_2 emission is better constrained, comparison with past carbon production inferred from petrological studies (Groppo et al. 2013, 2022; Rapa et al. 2017) will be possible in order to obtain a general overview of CO_2 emission from the largest Earth's orogen across timescales.

Finally, some key hydrothermal sites in central Nepal that have already proven their sensitivity to earthquake and deformation, through post-seismic and possibly pre-seismic observations (Girault et al. 2018c), should be used as a starting basis to set up an observatory network of the temporal variations of CO_2 emissions. Crustal fluids at hydrothermal systems in the Nepal Himalaya emerge as powerful assets for the characterization of the carbon budget of the Earth and the monitoring of current CO_2 emissions associated with the Himalayan seismic cycle.

9.7. References

Adhikari, L.B. (2021). Seismicity associated with the April 25, 2015 Gorkha earthquake in Nepal: Probing the Himalayan Seismic Cycle. Thesis, Université de Paris, Institut de Physique du Globe de Paris.

Adhikari, L.B., Gautam, U.P., Koirala, B.P., Bhattarai, M., Kandel, T., Gupta, R.M., Timsina, C., Maharjan, N., Maharjan, K., Dahal, T. et al. (2015). The aftershock sequence of the April 25th 2015 Gorkha-Nepal earthquake. *Geophysical Journal International*, 203, 2119–2124.

Adhikari, L.B., Laporte, M., Bollinger, L., Vergne, J., Lambotte, S., Koirala, B.P., Bhattarai, M., Timsina, C., Gupta, R.M., Wendling-Vazquez, N. et al. (In press 2022). Seismically active structures of the Main Himalayan Thrust revealed before, during and after the 2015 Mw 7.9 Gorkha earthquake in Nepal. *Geophysical Journal International*, ggac281. https://doi.org/210.1093/gji/ggac1281.

Ambraseys, N.N. and Douglas, J. (2004). Magnitude calibration of north Indian earthquakes. *Geophysical Journal International*, 159, 165–206.

Ambraseys, N.N. and Jackson, D. (2003). A note on early earthquakes in northern India and southern Tibet. *Current Science*, 84, 570–582.

Becker, J.A. (2005). Quantification of Himalayan metamorphic CO_2 fluxes: Impact on global carbon budgets, Corpus Christi College. Thesis, University of Cambridge.

Becker, J.A., Bickle, M.J., Galy, A., Holland, T.J.B. (2008). Himalayan metamorphic CO_2 fluxes: Quantitative constraints from hydrothermal springs. *Earth and Planetary Science Letters*, 265, 616–629.

Bonini, M. (2022). Can coseismic static stress changes sustain postseismic degassing? *Geology*, 50, 371–376. https://doi.org/10.1130/G49465.49461.

Byrdina, S., Revil, A., Pant, S.R., Koirala, B.P., Shrestha, P.L., Tiwari, D.R., Gautam, U.P., Shrestha, K., Sapkota, S.N., Contraires, S. et al. (2009). Dipolar self-potential anomaly associated with carbon dioxide and radon flux at Syabru-Bensi hot springs in central Nepal. *Journal of Geophysical Research*, 114, B10101.

Chiodini, G., Cardellini, C., Di Luccio, F., Selva, J., Frondini, F., Caliro, S., Rosiello, A., Beddini, G., Ventura, G. (2020). Correlation between tectonic CO_2 Earth degassing and seismicity is revealed by a 10-year record in the Apennines, Italy. *Science Advances*, 6, eabc2938.

Derry, L.A., Evans, M.J., Darling, R., France-Lanord, C. (2009). Hydrothermal heat flow near the Main Central Thrust, central Nepal Himalaya. *Earth and Planetary Science Letters*, 286, 101–109.

Evans, M.J., Derry, L.A., France-Lanord, C. (2004). Geothermal fluxes of alkalinity in the Narayani river system of central Nepal. *Geochemistry, Geophysics, Geosystems*, 5, Q08011.

Evans, M.J., Derry, L.A., France-Lanord, C. (2008). Degassing of metamorphic carbon dioxide from the Nepal Himalaya. *Geochemistry, Geophysics, Geosystems*, 9, Q04021.

Frondini, F., Cardellini, C., Caliro, S., Beddini, G., Rosiello, A., Chiodini, G. (2019). Measuring and interpreting CO_2 fluxes at regional scale: The case of the Apennines, Italy. *Journal of the Geological Society of London*, 176, 408–416.

Gajurel, A.P., France-Lanord, C., Huyghe, P., Guilmette, C., Gurung, D. (2006). C and O isotope compositions of modern fresh-water mollusc shells and river waters from Himalaya and Ganga plain. *Chemical Geology*, 233, 156–183.

Galetzka, J., Melgar, D., Genrich, J.F., Geng, J., Owen, S., Lindsey, E.O., Xu, X., Bock, Y., Avouac, J.-P., Adhikari, L. et al. (2015). Slip pulse and resonance of Kathmandu basin during the 2015 Mw 7.8 Gorkha earthquake, Nepal imaged with geodesy. *Science*, 349, 1091–1095.

Garzione, C.N., Quade, J., DeCelles, P.G., English, N.B. (2000). Predicting paleoelevation of Tibet and the Himalaya from O vs. altitude gradients in meteoric water across the Nepal Himalaya. *Earth and Planetary Science Letters*, 183, 215–229.

Geersen, J., Scholz, F., Linke, P., Schmidt, M., Lange, D., Behrmann, J.H., Völker, D., Hensen, C. (2016). Fault zone controlled seafloor methane seepage in the rupture area of the 2010 Maule earthquake, central Chile. *Geochemistry, Geophysics, Geosystems*, 17, 4802–4813.

Ghezzi, L., Iaccarino, S., Carosi, R., Montomoli, C., Simonetti, M., Paudyal, K.R., Cidu, R., Petrini, R. (2019). Water quality and solute sources in the Marsyangdi River system of Higher Himalayan range (West-Central Nepal). *Science of the Total Environment*, 677, 580–589.

Girault, F. (2011). Etude des flux de dioxyde de carbone et de radon dans l'Himalaya du Népal (in French). Thesis, University Paris Diderot (Paris VII).

Girault, F. and Perrier, F. (2014). The Syabru-Bensi hydrothermal system in central Nepal 2: Modeling and significance of the radon signature. *Journal of Geophysical Research: Solid Earth*, 119, 4056–4089.

Girault, F., Koirala, B.P., Perrier, F., Richon, P., Rajaure, S. (2009). Persistence of radon-222 flux during monsoon at a geothermal zone in Nepal. *Journal of Environmental Radioactivity*, 100,955–100,964.

Girault, F., Perrier, F., Crockett, R., Bhattarai, M., Koirala, B.P., France-Lanord, C., Agrinier, P., Ader, M., Fluteau, F., Gréau, C. et al. (2014a). The Syabru-Bensi hydrothermal system in central Nepal 1: Characterization of carbon dioxide and radon fluxes. *Journal of Geophysical Research: Solid Earth*, 119, 4017–4055.

Girault, F., Bollinger, L., Bhattarai, M., Koirala, B.P., France-Lanord, C., Rajaure, S., Gaillardet, J., Fort, M., Sapkota, S.N., Perrier, F. (2014b). Large-scale organization of carbon dioxide discharge in the Nepal Himalayas. *Geophysical Research Letters*, 41, 6358–6366.

Girault, F., Perrier, F., Przylibski, T.A. (2018a). Radon-222 and radium-226 occurrence in water: A review. *Geological Society London Special Publications*, 451, 131–154.

Girault, F., Koirala, B.P., Bhattarai, M., Perrier, F. (2018b). Radon and carbon dioxide around remote Himalayan thermal springs. *Geological Society London Special Publications*, 451, 155–181.

Girault, F., Adhikari, L.B., France-Lanord, C., Agrinier, P., Koirala, B.P., Bhattarai, M., Mahat, S.S., Groppo, C., Rolfo, F., Bollinger, L. et al. (2018c). Persistent CO_2 emissions and hydrothermal unrest following the 2015 earthquake in Nepal. *Nature Communications*, 9, 2956.

Girault, F., Viveiros, F., Silva, C., Thapa, S., Pacheco, J.E., Adhikari, L.B., Bhattarai, M., Koirala, B.P., Agrinier, P., France-Lanord, C. et al. (2022). Radon signature of CO_2 flux constrains the depth of degassing: Furnas volcano (Azores, Portugal) versus Syabru-Bensi (Nepal Himalayas). *Scientific Reports*, 12, 10837.

Grabczak, J. and Kotarba, M. (1985). Isotopic composition of the thermal waters in the central part of the Nepal Himalayas. *Geothermics*, 14, 567–575.

Groppo, C., Rolfo, F., Castelli, D., Connolly, J.A.D. (2013). Metamorphic CO_2 production from calc-silicate rocks via garnet-forming reactions in the CFAS–H2O–CO_2 system. *Contributions to Mineralogy and Petrology*, 166, 1655–1675.

Groppo, C., Rolfo, F., Frezzotti, M.L. (2022). CO_2 outgassing during collisional orogeny is facilitated by the generation of immiscible fluids. *Communications Earth & Environment*, 3, 13.

Gualandi, A., Avouac, J.-P., Galetzka, J., Genrich, J.F., Blewitt, G., Adhikari, L.B., Koirala, B.P., Gupta, R., Upreti, B.N., Pratt-Sitaula, B. et al. (2017). Pre- and post-seismic deformation related to the 2015, Mw7.8 Gorkha earthquake, Nepal. *Tectonophysics*, 714–715, 90–106.

Gutenberg, B. and Richter, C.F. (1949). *Seismicity of the Earth and Associated Phenomena*. Princeton University Press, NJ.

Ingebritsen, S.E. and Manning, C.E. (2010). Permeability of the continental crust: Dynamic variations inferred from seismicity and metamorphism. *Geofluids*, 10, 193–205.

Manga, M., Beresnev, I., Brodsky, E.E., Elkhoury, J.E., Elsworth, D., Ingebritsen, S.E., Mays, D.C., Wang, C.-Y. (2012). Changes in permeability caused by transient stresses: Field observations, experiments, and mechanisms. *Reviews of Geophysics*, 50, RG2004.

Manning, C.E. and Ingebritsen, S.E. (1999). Permeability of the continental crust: Implications of geothermal data and metamorphic systems. *Reviews of Geophysics*, 37, 127–150.

Marty, B., France-Lanord, C., Gajurel, A.P., Muralt, R., Vuatz, F., Le Fort, P. (1996). CO_2 and helium in mineral springs of Nepal and Mustang: A study of volatile release processes during the Himalayan tectonics. Report, Abstracts of the 11th Himalaya-Karakoram-Tibet Conference, Northern Arizona University, Flagstaff AZ.

Miller, S.M. (2020). Aftershocks are fluid-driven and decay rates controlled by permeability dynamics. *Nature Communications*, 11, 5787.

Perrier, F., Chitrakar, G.R., Froidefond, T., Tiwari, D., Gautam, U., Kafle, B., Trique, M. (2002). Estimating streaming potentials associated with geothermal circulation at the Main Central Thrust: An example from Tatopani-Kodari hot spring in central Nepal. *Journal of Nepal Geological Society*, 26, 17–27.

Perrier, F., Richon, P., Byrdina, S., France-Lanord, C., Rajaure, S., Koirala, B.P., Shrestha, P.L., Gautam, U.P., Tiwari, D.R., Revil, A. et al. (2009). A direct evidence for high carbon dioxide and radon-222 discharge in Central Nepal. *Earth and Planetary Science Letters*, 278, 198–207.

Rai, S.M., Bhattarai, T.N., Khatiwada, D. (2020). Hot water springs (thermal springs) in Nepal: A review on their location, origin, and importance. *Journal of Development Innovations*, 4, 24–42.

Ranjit, M. (2015). Geothermal energy update of Nepal. *Proceedings of the World Geothermal Congress*, 1–10.

Rapa, G., Groppo, C., Rolfo, F., Petrelli, M., Mosca, P., Perugini, D. (2017). Titanite-bearing calc-silicate rocks constrain timing, duration and magnitude of metamorphic CO_2 degassing in the Himalayan belt. *Lithos*, 292–293, 364–378.

Sapkota, S.N., Bollinger, L., Klinger, Y., Tapponnier, P., Gaudemer, Y., Tiwari, D. (2013). Primary surface ruptures of the great Himalayan earthquakes in 1934 and 1255. *Nature Geoscience*, 6, 71–76.

Singh, R., Kanwar, S.S., Jaggi, G.S., Kartha, K.N.K. (2004). Geochemistry of thermal springs from Bhutan Himalaya. *Journal of the Geological Society of India*, 64, 191–198.

Szeliga, W., Hough, S., Martin, S., Bilham, R. (2010). Intensity, magnitude, location, and attenuation in India for felt earthquakes since 1762. B. *Seismological Society of America*, 100, 570–584.

Tiwari, S.K., Rai, S.K., Bartarya, S.K., Gupta, A.K., Negi, M. (2016). Stable isotopes ($\delta^{13}C_{DIC}$, δD, $\delta^{18}O$) and geochemical characteristics of geothermal springs of Ladakh and Himachal (India): Evidence for CO_2 discharge in northwest Himalaya. *Geothermics*, 64, 314–330.

Zheng, G., Xu, S., Liang, S., Shi, P., Zhao, J. (2013). Gas emission from the Qingzhu River after the 2008 Wenchuan Earthquake, Southwest China. *Chemical Geology*, 339, 187–193.

Conclusion

Rodolphe CATTIN[1] and Jean-Luc EPARD[2]

[1] *University of Montpellier, France*

[2] *University of Lausanne, Switzerland*

In this third volume, we have presented some elements to understand better the scientific approach used by researchers to unravel many processes acting in the Himalayan dynamics. We have focused in particular on:

– the climatic effect of an orogeny;

– the action of denudation on mountain landscapes;

– the impact of climate warming on glaciers and glaciers outburst floods;

– the improvement of seismic hazard assessment;

– the evolution of the hydrothermal systems observed along the Himalayan arc.

Numerous field, laboratory and modeling studies have led to significant advances in understanding Himalayan dynamics over the past decades. Far from being a cylindrical geological structure associated with the collision between India and Eurasia, the Himalayan range is a complex structure whose many stages of evolution are still poorly understood. The studies of the lateral variations of the geological structures, the fluid circulations, the erosion processes and the tectonic-climate interactions remain open scientific questions.

Seismic risk is high in the Himalaya. Recent studies have allowed us to better characterize this region's seismicity, location and evolution. However,

Himalaya, Dynamics of a Giant 3,
coordinated by Rodolphe CATTIN and Jean-Luc EPARD.

much remains to be done. The implementation of dense networks, satellite observations and the reinforcement of building construction rules are all avenues to be explored to reduce this risk.

Last but not least, mountain environments are at the forefront of climate change, with widespread glacier retreat and a dramatic increase in the frequency of extreme events. The impact of global warming on the Himalaya is likely to have severe consequences. Landslides and flash floods can cause devastating damage to infrastructures, buildings and people. Prediction and warning systems are therefore essential to secure mountain livelihoods. Melting Himalayan glaciers will reduce water availability for irrigation and hydropower production, potentially putting food and energy security of tens of millions of people at risk, including those living in the plains.

Understanding the delicate balance that controls the dynamics of the Himalayan giant is therefore now, more than ever, a major challenge for the scientific community.

List of Authors

Lok Bijaya ADHIKARI
Department of Mines and Geology
Kathmandu
Nepal

Pierre AGRINIER
Paris Cité University
France

Awais AHMED
Geological Survey of Pakistan
Islamabad
Pakistan

Christoff ANDERMANN
GFZ
Potsdam
Germany

Mikaël ATTAL
University of Edinburgh
United Kingdom

Adnan Alam AWAN
Geological Survey of Pakistan
Islamabad
Pakistan

Mohd Farooq AZAM
Indian Institute of Technology Indore
Simrol
India

Laurent BOLLINGER
French Alternative Energies and
Atomic Energy Commission
Bruyères-le-Châtel
France

Fanny BRUN
University of Grenoble Alpes
France

Rodolphe CATTIN
University of Montpellier
France

Luca DAL ZILIO
ETH Zürich
Switzerland

Saptarshi DEY
Indian Institute of Technology
Gandhinagar
India

Yannick DONNADIEU
CEREGE
Aix-en-Provence
France

Jean-Luc EPARD
University of Lausanne
Switzerland

Matthieu FERRY
University of Montpellier
France

Frédéric FLUTEAU
Institut de physique du globe de Paris
Paris Cité University
France

Monique FORT
Paris Cité University
France

Christian FRANCE-LANORD
University of Nancy
Vandoeuvre-lès-Nancy
France

Ananta Prasad GAJUREL
Tribhuvan University
Kathmandu
Nepal

Eduardo GARZANTI
Università degli Studi di
Milano-Bicocca
Milan
Italy

Frédéric GIRAULT
Paris Cité University
France

Kaushal GNYAWALI
School of Engineering
University of British Columbia
Kelowna
Canada
and
Natural Hazards Section
Himalayan Risk Research Institute
Bhaktapur
Nepal

Vincent GODARD
CEREGE
Aix-en-Provence
France

Chiara GROPPO
Department of Earth Sciences
University of Turin
and
Institute of Geoscience and Geohazard
Turin
Italy

György HETÉNYI
Institute of Earth Sciences
University of Lausanne
Switzerland

Ramperu JAYANGONDAPERUMAL
Wadia Institute of Himalayan Geology
Dehra Dun
India

Romain JOLIVET
Ecole Normale Supérieure
Paris
France

François JOUANNE
Savoie Mont Blanc University
Le Bourget-du-Lac
France

Yann KLINGER
Paris Cité University
France

Guillaume LE HIR
Institut de physique du globe de Paris
Paris Cité University
France

Romain LE ROUX-MALLOUF
EDF
Aix-en-Provence
France

Rémi LOSNO
Paris Cité University
France

Maarten LUPKER
Independent researcher
Bern
Switzerland

Sudan Bikash MAHARJAN
International Centre for Integrated Mountain Development
Kathmandu
Nepal

Manzoor A. MALIK
University of Jammu
India

Odin MARC
Paul Sabatier University
Toulouse
France

Sylvain MICHEL
Ecole Normale Supérieure
Paris
France

Jean-Louis MUGNIER
Savoie Mont Blanc University
Le Bourget-du-Lac
France

Naveed MUNAWAR
Geological Survey of Pakistan
Islamabad
Pakistan

Santosh NEPAL
International Centre for Integrated Mountain Development
Kathmandu
Nepal

Kabi Raj PAUDYAL
Tribhuvan University
Kathmandu
Nepal

Frédéric PERRIER
Paris Cité University
France

Martin ROBYR
University of Lausanne
Switzerland

Franco ROLFO
Department of Earth Sciences
University of Turin
and
Institute of Geoscience and Geohazard
Turin
Italy

Anta-Clarisse SARR
CEREGE
Aix-en-Provence
France

Wolfgang SCHWANGHART
University of Potsdam
Germany

Victoria STEVENS
Geological Sciences Department
University of Cape Town
Rondebosch
South Africa
and
Nanyang Technological University
Singapore

Shiba SUBEDI
Institute of Earth Sciences
University of Lausanne
Switzerland
and
Seismology at School in Nepal
Pokhara
Nepal

Shashi TAMANG
University of Turin
Italy
and
Paris Cité University
France
and
Tribhuvan University
Kathmandu
Nepal

Delphine TARDIF
LSCE
Gif-sur-Yvette
and
Institut de Recherche pour le Développement
Montpellier
France

Sandeep THAPA
Paris Cité University
France

Rasmus C. THIEDE
Christian-Albrecht University
Kiel
Germany

Bishal Nath UPRETI
Nepal Academy of Science and Technology
Kathmandu
Nepal

Jérôme VAN DER WOERD
University of Strasbourg
France

Riccardo VASSALLO
Savoie Mont Blanc University
Le Bourget-du-Lac
France

Georg VEH
University of Potsdam
Germany

Jérôme VERGNE
University of Strasbourg
France

Patrick WAGNON
University of Grenoble Alpes
France

Index

E, F

G, H

I, J

U, W

Summary of Volume 1

Part 2. Along Strike Variations

Chapter 4. Seismological Imaging and Current Seismicity of the Himalayan Arc

György HETÉNYI, Jérôme VERGNE, Laurent BOLLINGER, Shiba SUBEDI, Konstantinos MICHAILOS and Dowchu DRUKPA

Chapter 5. Gravity Observations and Models Along the Himalayan Arc

Rodolphe CATTIN, György HETÉNYI, Théo BERTHET and Jamyang CHOPHEL

Summary of Volume 2

Part 2. Greater Himalayan Crystalline Complex

Chapter 4. High-Pressure and Ultra-High-Pressure Units in the Himalaya

Julia DE SIGOYER and Stéphane GUILLOT

Chapter 5. The Greater Himalayan Sequence – Tectonic, Petrographic and Kinematic Evolution of the Metamorphic Core Zone of the Himalayan Orogeny

Martin ROBYR, Rodolfo CAROSI, Salvatore IACCARINO and Chiara MONTOMOLI

Printed and bound by CPI Group (UK) Ltd, Croydon, CR0 4YY
13/09/2026
14960999-0005